D1809006

# Turbo Pascal Programs
# for Scientists and Engineers

# Turbo Pascal® Programs
# for Scientists and Engineers

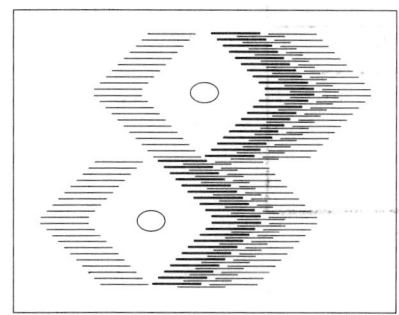

Alan R. Miller

San Francisco • Paris • Düsseldorf • London    SYBEX®

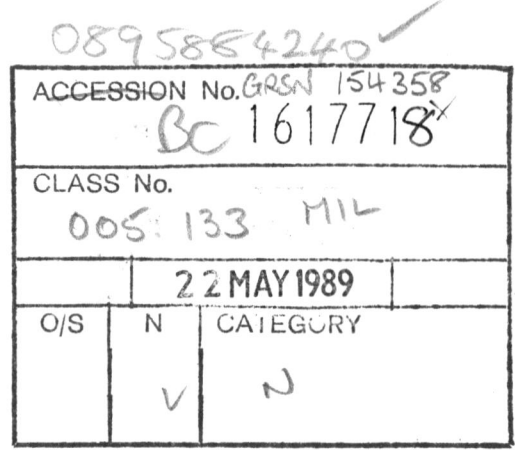

0895884240 ✓

ACCESSION No. GRSN 154358
BC 161771 8 ×

CLASS No.
005: 133 MIL

2 2 MAY 1989

| O/S | N | CATEGORY |
|---|---|---|
|  | V | N |

Cover art by Thomas Ingalls + Associates
Cover photography by David Bishop

**IBM PC** and **PC-DOS** are trademarks of International Business Machines Corporation.
**Grammatik** is a trademark of Aspen Software Company.
**Sidekick** and **Superkey** are trademarks and **Turbo Pascal** is a registered trademark of Borland
 International.
**WordStar** is a trademark of MicroPro International Corporation.

SYBEX is a registered trademark of SYBEX, Inc.

SYBEX is not affiliated with any manufacturer.

Every effort has been made to supply complete and accurate information. However,
SYBEX assumes no responsibility for its use, nor for any infringements of patents or
other rights of third parties which would result.

Copyright©1987 SYBEX Inc., 2021 Challenger Drive #100, Alameda, CA 94501.
World rights reserved. No part of this publication may be stored in a retrieval system, transmitted, or reproduced in any way, including but not limited to photocopy,
photograph, magnetic or other record, without the prior agreement and written per-
mission of the publisher.

Library of Congress Card Number:
ISBN 0-89588-518-2
Manufactured in the United States of America
10 9 8 7 6 5 4 3 2

# *Acknowledgments*

I am sincerely grateful to Marilyn Smith, editor of the manuscript, for her many helpful suggestions. Robert Campbell reviewed the manuscript and programs from the technical aspect and made valuable comments. Rudolph Langer and Douglas Hergert gave helpful early direction. I would also like to thank Ashok Singh for reviewing the mathematical aspects of the manuscript. Other SYBEX staff members who made contributions are Olivia Shinomoto, word processing coordinator; Michelle Hoffman, graphics technician; Stephanie Bower, proofreader; and Sharon Leong, designer.

Alan R. Miller
Socorro, New Mexico 87801
April 19, 1987

# *Contents*

## *Introduction*                                                                 *xix*

## *Chapter 1*                                                                      *1*
### *Evaluating the Turbo Pascal Compiler*

Turbo Pascal Compilers                                              1

Precision and Range of Floating-Point Operations      2

    *Testing Floating-Point Operations*                       3
    *Comparing the Compilers*                                 4

Other Turbo Pascal Functions                                   5

    *Testing Functions*                                       5
    *The Arc Tangent Function*                                7
    *The Arc Sine and Arc Cosine Functions*                   8
    *Testing the Arc Sine and Arc Tangent Functions*         9

Using External Files                                               10

Summary                                                            11

Exercises                                                          11

## *Chapter 2*                                                                     *15*
### *Statistical Tools*

The Mean                                                           15

    *Dispersion from the Mean*                                16

The Standard Deviation                                             18

    *Calculation of the Standard Deviation*                   19

Calculating the Mean and Standard Deviation                        20

Random Numbers and Random-Number
  Generators        23

*Evaluating the Random-Number Generator*    23
*Using π to Produce Random Numbers*    25
*Gaussian Random Numbers*    25
*A Gaussian Random-Number Generator*    26

Summary    28

Exercises    28

## Chapter 3    *31*

## Vector and Matrix Operations

Scalars and Arrays    31

Vectors    32

*Vector Arithmetic*    33
*Strings*    35

Matrices    36

*Matrix Arithmetic*    37

Programming Matrix Multiplication    40

Determinants    44

Calculating Determinants    45

Inverse Matrices and Matrix Division    46

Summary    47

Exercises    47

## Chapter 4    *51*

## Simultaneous Solution of Linear Equations

Linear Equations and Simultaneous Solutions    52

Solution by Cramer's Rule    53

*Linear Dependence*    54
*Example: A Direct-Current Electrical Circuit*    55

A More Elegant Use of Cramer's Rule      57

    *Running the Cramer's Rule Program*      60

Solution by Gauss Elimination      60

    *The Steps of the Gauss Elimination Method*      61

    *Improving the Accuracy of the Gauss Elimination*
        *Method*      63

    *Programming the Gauss Elimination Method*      63

    *Running the Gauss Elimination Program*      67

Solution by Gauss-Jordan Elimination      68

    *Details of the Gauss-Jordan Elimination Method*      69

    *Programming the Gauss-Jordan Elimination Method*      70

    *Running the Gauss-Jordan Elimination Program*      75

Multiple Constant Vectors and Matrix Inversion      75

An Alternative Method for Programming Gauss-
    Jordan Elimination      77

    *Running the Alternative Version of the Gauss-Jordan*
        *Elimination Program*      77

Ill-Conditioned Equations      84

    *The Hilbert Matrix*      84

    *Running the Hilbert Matrix Program*      87

Reading Coefficients from a Disk File      89

    *Running the Disk-File Program*      89

A Simultaneous Best Fit      92

    *Programming the Best-Fit Solution*      93

    *Running the Best-Fit Program*      96

Equations with Complex Coefficients      97

    *Example: An Alternating-Current Electrical Circuit*      98

    *Programming Simultaneous Equations with Complex*
        *Coefficients*      101

    *Running the Complex-Coefficient Program*      104

The Gauss-Seidel Iterative Method      105

    *Programming the Gauss-Seidel Method*      107

    *Running the Gauss-Seidel Program*      111

Summary      112

Exercises      112

## Chapter 5          *115*
### *Development of a Curve-Fitting Program*

The Main Program ............ 116
   *First Version Using the Built-In Random Function* ... 116
   *The Scattering Algorithm* ... 118
   *Running the Main Program* ... 119

Programming a Plotter Routine ... 120
   *Adding the Plotter Routine to the Main Program* ... 120
   *Running the New Version* ... 124

A Simulated Curve Fit ... 125
   *Running the Plotter Routine with Two Curves* ... 126

The Curve-Fitting Algorithm ... 128
   *The Curve-Fitting Procedure* ... 131
   *Running the Curve-Fitting Program* ... 132

The Correlation Coefficient ... 132
   *Programming Correlation Coefficient Calculations* ... 134
   *Running the Finished Program* ... 138

Summary ... 138

Exercises ... 139


## Chapter 6          *143*
### *Sorting*

Handling Experimental Data ... 143

Testing the Sorting Routines ... 144
   *The Test and Swap Routines* ... 146

The Bubble Sort ... 147
   *Programming the Bubble Sort* ... 147
   *Testing the Bubble Sort* ... 148

The Shell Sort ... 148
   *Programming the Shell Sort* ... 149
   *Testing the Shell Sort* ... 149

The Quick Sort ... 150
   *Programming the Quick Sort* ... 151

*Testing the Quick Sort*                                       151

Adding a Sorting Routine to the Curve-Fitting
   Program                                                     152
   *Revising the Curve-Fitting Program*                       153
   *Revising the Shell Sort Routine*                          153
   *Running the Revised Curve-Fitting Program*                154

Sorting Data Stored on Disk                                   155
   *Programming the Record Sorting Routine*                   155
   *Running the Record Sorting Program*                       157

Summary                                                       157

Exercises                                                     158

## *Chapter 7*                                              *161*

## *Generalized Least-Squares Curve Fitting*

A Parabolic Curve Fit                                         162
   *Programming a Least-Squares Curve Fit for a
      Parabola*                                               163
   *Running the Program*                                      167

Fitting Curves to Other Equations                             168
   *A Direct Solution*                                        169
   *Programming the Matrix Approach to Curve Fitting*         170
   *Running the Parabolic Curve-Fitting Program*              174

Adjusting the Order of the Polynomial                         175
   *Comparing Runs Using Different Polynomial Orders*         178

The Heat-Capacity Equation                                    180

The Vapor Pressure Equation                                   183

The Equation for the Properties of Superheated
   Steam                                                      186
   *A Three-Variable Equation with a Nonlinear
      Coefficient*                                            187
   *An Equation of State for Steam*                           188
   *Comparing Runs of the Steam Properties Program*          191

Summary                                                       194

Exercises                                                     195

## Chapter 8 — 199
### Solution of Equations by Newton's Method

Formulating Newton's Method                                    200
  *Approximating Values*                                       202
Programming Newton's Method                                    205
  *Tolerance*                                                  206
  *Generalizing Procedure Calls*                               207
  *Running the Newton's Method Program*                        207
  *Adding User Input for the First Approximation*              208
  *Running the Program to Find the Second Root*                209
  *A Test for Zero Slope*                                      209
  *Running the Program with the Slope Test*                    210
  *Failure to Converge*                                        212
Programming the Solution to Other Equations                    214
  *A Function with Many Roots*                                 215
Solving the Vapor Pressure Equation                            216
Summary                                                        216
Exercises                                                      217

## Chapter 9 — 221
### Numerical Integration

The Definite Integral                                          222
The Trapezoidal Method                                         223
  *Programming the Trapezoidal Method*                         225
  *Programming an Improved Trapezoidal Method*                 227
  *Running the Improved Trapezoidal Method Program*            228
Simpson's Method                                               229
  *Programming Simpson's Method*                               230
  *Using an Exponential Function with the Simpson's
    Method Program*                                            232
  *Using a Periodic Function with the Simpson's Method
    Program*                                                   232

The Romberg Method     234

*Programming the Romberg Method*     235

*First Run of the Romberg Method Program*     237

*Using an Exponential Function with the Romberg Method Program*     237

*Using a Periodic Function with the Romberg Method Program*     239

Solving Functions that Become Infinite at One Limit     240

*Programming Adjustable Panels for an Infinite Function*     240

*Running the Adjustable Panel Program*     243

Summary     243

Exercises     244

## *Chapter 10*     *247*

## *Nonlinear Curve-Fitting Equations*

Linearizing the Rational Function     248

Fitting the Clausing Factor to the Rational Function     248

*Running the Clausing Factor Fitting Program*     252

Linearizing the Exponential Equation     253

*Programming an Exponential Curve Fit*     256

*Running the Exponential Curve-Fitting Program*     256

Direct Solution of the Exponential Equation     257

*Calculating the SRS*     258

*Eliminating Coefficient A*     258

*Applying Newton's Method*

*Programming a Nonlinearized Exponential Curve Fit*     259

*Running the Nonlinearized Exponential Curve-Fitting Program*     263

Summary     264

Exercises     264

## Chapter 11    *267*

### *Advanced Applications*

The Normal and Cumulative Distribution
Functions                                    268
   *The Standard Deviation*                   269

The Gaussian Error Function                  271
   *Diffusion in a One-Dimensional Slab*     271

Evaluating the Gaussian Error Function Using
Simpson's Method                             273
   *Calculating the Error Function Using Simpson's*
   *Method*                                  273

Evaluating the Gaussian Error Function Using an
Infinite Series Expansion                    275
   *Running the Infinite Series Expansion Program*   277
   *Using the Infinite Series Expansion Program*     277

The Complement of the Error Function         278
   *Evaluating the Complement of the Error Function*   278
   *Running the Error Function Complement Program*     281

The Gamma Function                           281
   *Evaluating the Gamma Function*           283
   *Running the Gamma Function Testing Program*   284

Bessel Functions                             286
   *Programming Bessel Functions of the First Kind*    286
   *Programming Bessel Functions of the Second Kind*   290

Summary                                      293

Exercises                                    294

## Appendix A    *297*

### *Reserved Words and Standard Identifiers*

Reserved Words                               297

Standard Identifiers                         298

## *Appendix B*                                    *301*

### *Summary of Turbo Pascal*

Standard Character Set                        301

Variable Names (Identifiers)                  301

Numbers                                       302

Comments                                      302

Operations                                    302
   *Integer Operations*        302
   *Real Operations*           303
   *Boolean Operations*        303
   *Relational Operations*     303

Syntax                                        303
   *Program Heading*           303
   *Constant Definition*       303
   *Variable Definition*       304
   *Assignment Statements*     304
   *The Compound Statement*    304
   *Procedure Definition*      305
   *Function Definition*       305
   *Placement of Semicolons*   306

Conditional Statements                        306
   *The IF-THEN Statement*      306
   *The IF-THEN-ELSE Statement* 306

Iterative Statements                          307
   *The WHILE-DO Statement*     307
   *The REPEAT-UNTIL Statement* 307
   *The FOR-TO-DO Statement*    307
   *The FOR-DOWNTO-DO Statement* 307

Transfer of Control Statements                308
   *The Exit Statement*         308
   *The Halt Statement*         308

Input and Output                              308
   *Input Procedures*           308
   *Output Procedures*          309
   *Formatting Numeric Output*  309
   *Data Types*                 310

*Scalar Types*     310
*Subrange Types*     310
*Array Types*     311
*Referencing Array Types*     311
*Set Types*     312
*Set Operations*     312
File Types     312

## Appendix C     *315*

### Answers to Exercises

## Index     *316*

# *Introduction*

The purpose of this book is twofold: to help the reader develop a proficiency in the use of Turbo Pascal, and to provide a library of programs useful for solving problems frequently encountered in scientific and engineering applications.

The ideas and material for this book have been developed during my experience teaching sophomore, junior, and senior engineering students since 1967. Over the years, the computer language I have used for these classes has changed from Fortran to BASIC, to what is perhaps the ideal language for the classroom—Pascal. However, the usual Pascal compiler has serious limitations that restrict its usefulness. Fortunately, Turbo Pascal has overcome these problems.

The programs in this book will prove valuable to the practicing scientist or engineer. The material is also suitable for students of junior- or senior-level engineering courses in numerical methods. The reader should have a working knowledge of an applications language such as Pascal, Fortran, or BASIC. Experience with vector operations and with differential and integral calculus will be helpful.

The distinctive features of Pascal lead to clear and elegant programming practices. Two of these features that are particularly valuable in scientific programs are long variable names and block structure. When identifiers such as Sum_X_Squared can be chosen in preference to shorter symbols, the resulting source code is easier to comprehend. Pascal also offers a greater variety of iterative statements than those available with BASIC and Fortran. The statement

```
FOR i := 1 TO n DO
```

corresponds to the loop in BASIC

```
FOR

...

NEXT
```

and the

    DO

    ...

    CONTINUE

loop in Fortran. In Pascal, this statement is complemented by the other iterative constructions

    WHILE

    ...

    DO

and

    REPEAT

    ...

    UNTIL

The major features of Pascal are summarized in the second edition of *Pascal User Manual and Report*, by Kathleen Jensen and Niklaus Wirth (Springer-Verlag, 1974); additional details can be found in *Introduction to Turbo Pascal*, by Douglas S. Stivison (SYBEX, 1985), and in the Turbo Pascal manual. Several desirable features, omitted from the original definition of Pascal, are included in Turbo Pascal. One such feature is the dynamic string type. Another is the inclusion of a random-number generator, a feature discussed in Chapter 2.

One limitation of Turbo Pascal is that it does not provide the ability to use procedure and function names as parameters to other procedures. This problem is discussed in Chapter 8, in the section titled Generalizing Procedure Calls.

Throughout this book, I have used features common to other high-level languages (such as Fortran, BASIC, and Algol) rather than the more sophisticated techniques specific to Turbo Pascal. For example, the programs use arrays instead of records. As a consequence, it is easy to convert the algorithms presented here into other languages.

Readers who are primarily interested in the Turbo Pascal programs developed in this book can locate the program listings in text or refer to the program index. However, this book is designed to be read from beginning to end. Each chapter discusses and develops tools that will be used again in subsequent chapters. The mathematical algorithms of each program are described in detail before the program itself is presented, and sample program output is illustrated. The following brief descriptions summarize the contents of each chapter.

- Chapter 1, Evaluation of the Turbo Pascal Compiler, demonstrates the range and precision of the Turbo Pascal and Turbo-87 Pascal compilers. The chapter also discusses the logarithmic, exponential, and trigonometric functions of Turbo Pascal and presents an improved arc tangent function.

- Chapter 2, Statistical Tools, discusses some fundamental statistical algorithms and presents a program for implementing them. Procedures for generating and testing both uniformly and Gaussian distributed random numbers are also given.

- Chapter 3, Vector and Matrix Operations, summarizes the operations of vector and matrix arithmetic, including dot product, cross product, matrix multiplication, and matrix inversion. Two important programs are developed—one for matrix multiplication and another for calculating determinants.

- Chapter 4, Simultaneous Solution of Linear Equations, presents programs for solving simultaneous equations by Cramer's rule, the Gauss elimination method, the Gauss-Jordan elimination method, and the Gauss-Seidel method. The chapter also examines ill conditioning by observing a program that generates Hilbert matrices and develops a program for solving equations with complex coefficients. Another program solves simultaneous equations when the coefficients are stored on disk.

- Chapter 5, Development of a Curve-Fitting Program, is the first of a series of chapters on curve fitting. Using a top-down program development approach, a linear least-squares curve-fitting program is written and discussed. The program includes procedures to simulate data, plot curves, compute the fitted curve, and supply the correlation coefficient.

- Chapter 6, Sorting, describes and compares three Turbo Pascal sorting routines, including a bubble sort, a Shell sort, and a recursive quick sort. A sorting routine is incorporated into the curve-fitting program of Chapter 5 to enable the program to handle real experimental data. Then a program is written to sort records that are stored on disk.

- Chapter 7, Generalized Least-Squares Curve Fitting, extends the curve-fitting program to general polynomial equations and fits curves for three typical calculations: heat capacity, vapor pressure, and properties of superheated steam.

- Chapter 8, Solution of Equations by Newton's Method, presents a series of programs that use Newton's method for finding the roots of equations. This tool will be used again in Chapter 10 for nonlinear curve fitting.

- Chapter 9, Numerical Integration, develops programs for three different integration techniques: the trapezoidal method, Simpson's method, and the Romberg method. Simpson's method will be used in Chapter 11 for evaluating the Gaussian error function.

- Chapter 10, Nonlinear Curve-Fitting Equations, discusses curve-fitting algorithms for the rational function and the exponential function. The Clausing factor and the diffusion equation are used as examples.

- Chapter 11, Advanced Applications, addresses several advanced topics in programming for mathematical applications. This chapter summarizes and expands upon a number of the concepts presented earlier in the book.

For the reader who is approaching Turbo Pascal for the first time, a summary of its syntax, standard functions, and reserved words is included in the appendices. The real educational experience of this book, however, will be gained by carefully working through the programs themselves.

The Pascal programs and text of book were developed on a PC's Limited 286-12 microcomputer fitted with a math coprocessor. The operating system was PC-DOS version 3.2. The source programs were written with MicroPro's WordStar text editor and then compiled with Turbo Pascal. I printed the Turbo Pascal source programs with a Brother HR-12 daisywheel printer. The listings were then photostated and incorporated directly into the book. Consequently, the Turbo Pascal source programs have not been retyped.

The manuscript was created and edited with WordStar. The spelling was checked with WordStar's CorrectStar, a spelling checker, and Grammatik, a syntax checker. Computer displays shown in the figures were created with the HOTSHOT program by Symsoft. This resident program copies the video screen to a laser printer without altering the screen. The manuscript was submitted to SYBEX on floppy disks for final editing and electronic transmission to the photocomposer.

# *Evaluating the Turbo Pascal Compiler*

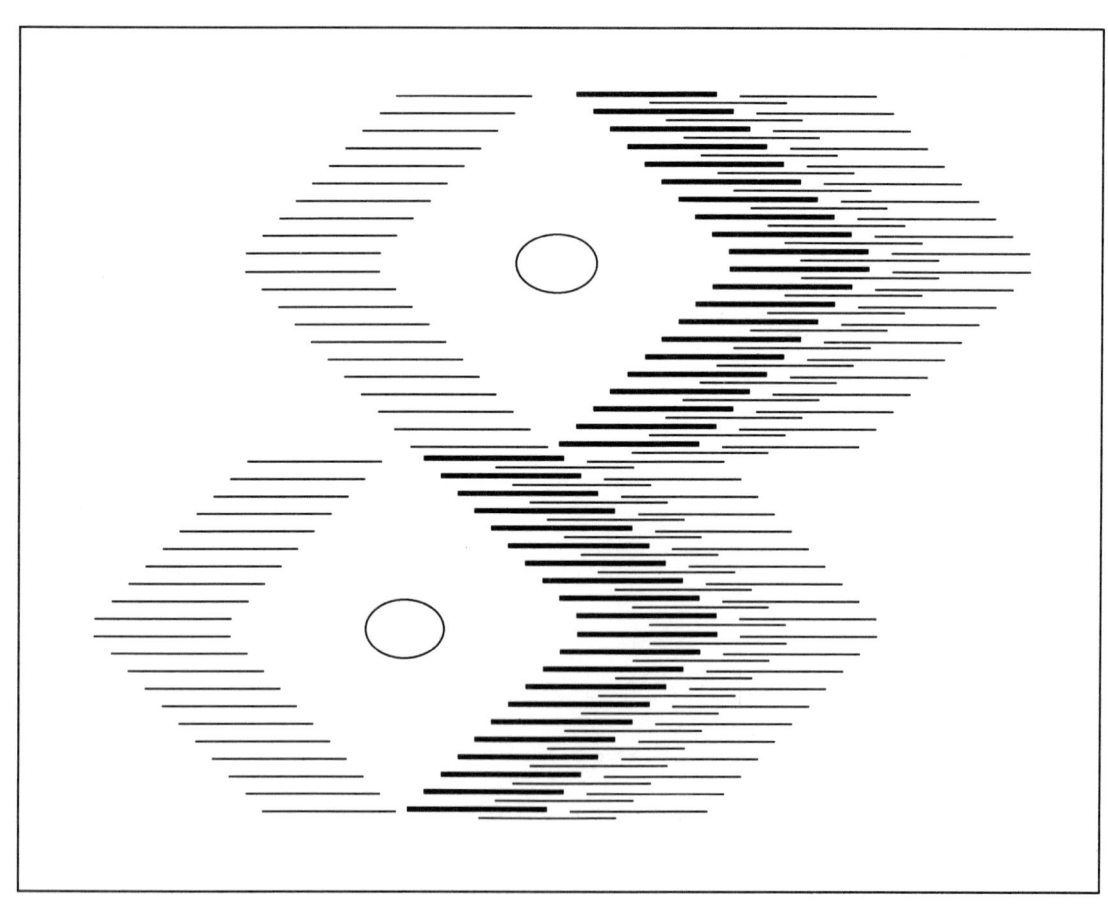

*Chapter* **1**

To understand the results of a Turbo Pascal program, you must be familiar with the limitations of the compiler that you are using. This is particularly true with scientific applications programs such as those presented in this book. In this first chapter, we will develop some tools for evaluating the precision and range of your compiler. We will also develop some useful functions that are not provided with the Turbo Pascal compiler.

Before we begin the evaluation, however, let us consider what a compiler is and how it can help you with your work. A computer can only manipulate numbers; it cannot work with mathematical expressions directly. You need to use a compiler to translate your mathematical equations and instructions into numbers that the computer can understand. First you write your Turbo Pascal instructions as a sequence of statements; then you run the compiler to convert them into machine language.

## *Turbo Pascal Compilers*

The regular Turbo Pascal compiler (called TURBO.COM) uses software to calculate floating-point operations. It has a precision of 11 decimal digits and a dynamic range of $10^{-38}$ to $10^{+38}$.

If you have a numeric coprocessor (8087 or 80287) in your computer, you should run the Turbo-87 Pascal compiler (called TURBO-87.COM) instead of the regular Turbo Pascal because this version uses the coprocessor for floating-point operations. The Turbo-87 compiler gives 15 digits of precision, and it has a dynamic range of $10^{-308}$ to $10^{+308}$. Furthermore, it calculates orders of magnitude faster than the regular Turbo Pascal compiler.

If you do not have a coprocessor, you should consider purchasing one. The cost of the integrated circuit ranges from \$100 to \$300, and it can be installed in a few minutes. On the other hand, you can use the regular Turbo Pascal compiler for all the programs in this book (both versions use binary arithmetic, which is suitable for scientific applications). However, without the coprocessor, the calculations will take longer and will not be as precise.

A third version of Turbo Pascal is designed for business applications. It performs all operations with software using decimal (BCD) arithmetic. This version gives 18 digits of precision and has a dynamic range of $10^{-63}$ to $10^{+63}$. It is the slowest of the three versions and does not provide scientific functions such as Sin and Exp. Therefore, the BCD version is not appropriate for the examples in this book, and we will not discuss it here.

## Precision and Range of Floating-Point Operations

Some of the programs in this book are sensitive to the *precision* and *dynamic range* of the Turbo Pascal floating-point operations. For example, several programs include an algorithm that generates a series of terms, ending when a particular term is smaller than a relative tolerance. The formula is

Term < Tol * Sum

where Term is the value of the new term, Sum is the total of the terms, and Tol is an arbitrarily small number known as the *tolerance*.

The value chosen for the tolerance must be within the accuracy of the floating-point operations. Otherwise, the summation step will never stop. Suppose, for example, that the floating-point operations are calculated to a precision of six significant figures. Then, the tolerance must be set to a value that is larger than $10^{-6}$. The dynamic range of the exponent is a separate matter.

The following section presents a program that tests the precision and range of Turbo Pascal. We will investigate output from the regular Turbo

Pascal compiler and the Turbo-87 version to evaluate the accuracy of their mantissa and exponent calculations.

## Testing Floating-Point Operations

The program given in Listing 1.1 can determine the precision and the dynamic range of Turbo Pascal. Enter the program using the file name TEST.PAS, and then run it.

Note that the output from the unaltered programs in this book will be displayed in black and white, even if you have a color monitor. Those of you with an enhanced graphics adapter (EGA) or a color/graphics adapter (CGA) can include the lines

```
TextColor (White);
TextBackground (Blue);
```

at the beginning of each program. The combination of white letters on a field of blue makes the display more pleasing and easier to read. Another pleasing combination is

```
TextColor (Black);
TextBackground (Cyan);
```

```
PROGRAM Test;
  { test range of floating-point numbers }
  { single-precision version }
{USES Crt;}  { Use for versions 4 and later }

VAR
  I: Integer;
  X: Real;

BEGIN
  TextColor(White); { for color screens }
  TextBackground(Blue); { for color screens }
  WriteLn;
  X := 1.0E-4 / 3.0;
  FOR I := 1 TO 18 DO
    BEGIN
      Write( X);
      X := 0.1 * X;
      WriteLn('     ', X);
      X := 0.1 * X
    END
END.
```

**Listing 1.1:** Testing floating-point operations.

which gives black letters on a field of blue-green.

The program to test the compiler's floating-point operations begins with the initial value of $X$, obtained by dividing $10^{-4}$ by 3. Then successively smaller and smaller values of $X$ are calculated and displayed on the video screen. Each succeeding value is obtained from the product of 0.1 and the previous value. The process continues until 18 values have been printed. We will use this program to demonstrate the accuracy and range of the two Turbo Pascal compilers.

## *Comparing the Compilers*

We choose the initial mantissa to be one-third, a repeating fraction that cannot be precisely represented by a floating-point number. Successive multiplications will show the extent of the roundoff error. Many compilers employ a 32-bit binary floating-point number that uses 3 bytes for the mantissa and 1 byte for the exponent. This usually produces six or seven significant figures of precision and a dynamic range of $10^{+38}$ to $10^{-38}$. However, the Turbo Pascal compilers are much more precise. The results from the regular Turbo Pascal compiler, shown in Figure 1.1, demonstrate a precision of 11 digits.

If you have a coprocessor installed in your computer and a copy of Turbo-87 Pascal, you can test it with the program given in Listing 1.1.

```
3.3333333333E-05     3.3333333333E-06
3.3333333333E-07     3.3333333333E-08
3.3333333333E-09     3.3333333333E-10
3.3333333333E-11     3.3333333333E-12
3.3333333333E-13     3.3333333333E-14
3.3333333333E-15     3.3333333333E-16
3.3333333333E-17     3.3333333333E-18
3.3333333333E-19     3.3333333333E-20
3.3333333333E-21     3.333333332E-22
3.333333332E-23      3.333333332E-24
3.333333332E-25      3.333333332E-26
3.333333332E-27      3.333333332E-28
3.333333332E-29      3.333333332E-30
3.333333332E-31      3.333333332E-32
3.333333332E-33      3.333333332E-34
3.333333332E-35      3.333333332E-36
3.333333332E-37      3.333333332E-38
3.333333332E-39      0.0000000000E+00
```

**Figure 1.1:** Precision test of the regular Turbo Pascal compiler.

However, change the eighth line to

X : 1.0E – 290/3.0;

The results should look like those shown in Figure 1.2. A comparison of Figures 1.1 and 1.2 shows that the Turbo-87 compiler gives four more figures of precision and a much larger dynamic range ($10^{-308}$). As we have seen, this version also performs the computations many times faster than the regular version.

```
3.33333333333333E-291        3.33333333333333E-292
3.33333333333333E-293        3.33333333333333E-294
3.33333333333333E-295        3.33333333333333E-296
3.33333333333333E-297        3.33333333333333E-298
3.33333333333333E-299        3.33333333333333E-300
3.33333333333334E-301        3.33333333333334E-302
3.33333333333334E-303        3.33333333333334E-304
3.33333333333334E-305        3.33333333333334E-306
3.33333333333334E-307        3.33333333333334E-308
0.33333333333333E-308        0.03333333333333E-308
0.03333333333333E-308        0.00033333333333E-308
0.00333333333333E-308        0.00003333333333E-308
0.00003333333333E-308        0.00000333333333E-308
0.00000033333333E-308        0.00000003333333E-308
0.00000003333333E-308        0.00000000333333E-308
0.00000000033333E-308        0.00000000033333E-308
0.00000000003333E-308        0.00000000003333E-308
0.00000000000033E-308        0.00000000000003E-308
0.00000000000000E-308        0.00000000000000E-308
0.00000000000000E+000        0.00000000000000E+000
```

**Figure 1.2:** Precision test of the Turbo-87 Pascal compiler.

# *Other Turbo Pascal Functions*

In this section, we will look at a routine designed to test other Turbo Pascal functions. We will also write some trigonometric functions that are not included with the Turbo Pascal compilers. Finally, we will examine the problematic Turbo Pascal ArcTan function.

## *Testing Functions*

We can test certain functions by pairing them with their inverses. For example, we can take the ArcTan of the ratio of Sin/Cos. (The Tan function is not included in Turbo Pascal.) As another example, we can take the Log of the Exp, as demonstrated by the program given in

Listing 1.2. This program should produce the original values. To try this out, create a file named TLOG.PAS, enter the program, and run it. The result should look like Figure 1.3 for the Turbo-87 version of Turbo Pascal. Notice that the results agree with the original numbers to all 15 digits.

```
PROGRAM Tlog;
 { test ln and exp }

VAR
  I: Integer;
  X, Y: Real;

BEGIN
  X := 1.0E-4 / 0.3;
  FOR I := 1 TO 10 DO
    BEGIN
      Y := Ln(X);
      WriteLn(' X =', X,', Exp(Ln) =', Exp(Y));
      X := 0.5 * X
    END
END.
```

**Listing 1.2:** Testing the Ln and Exp functions.

```
x =  3.33333333333333E-004, exp(ln) =  3.33333333333334E-004
x =  1.66666666666667E-004, exp(ln) =  1.66666666666667E-004
x =  8.33333333333333E-005, exp(ln) =  8.33333333333334E-005
x =  4.16666666666667E-005, exp(ln) =  4.16666666666667E-005
x =  2.08333333333333E-005, exp(ln) =  2.08333333333333E-005
x =  1.04166666666667E-005, exp(ln) =  1.04166666666667E-005
x =  5.20833333333333E-006, exp(ln) =  5.20833333333333E-006
x =  2.60416666666667E-006, exp(ln) =  2.60416666666667E-006
x =  1.30208333333333E-006, exp(ln) =  1.30208333333333E-006
x =  6.51041666666667E-007, exp(ln) =  6.51041666666667E-007
```

**Figure 1.3:** Results of the TLOG program.

### *The Arc Tangent Function*

Turbo Pascal includes the standard functions Sqrt, Sqr, Exp, Ln, Sin, Cos, and ArcTan. Unfortunately, there is a problem with the ArcTan function: it has not been programmed uniformly for negative arguments. Turbo Pascal returns an angle in the fourth quadrant when the argument is negative. However, some compilers return an angle in the second quadrant (90° to 180°), and others return a value in the fourth quadrant (0° to −90°). Furthermore, since there is only a single argument to Turbo Pascal's ArcTan function, it is not possible to distinguish between angles in the first quadrant and those in the third quadrant. Instead of trying to use this function, let's write our own version.

The arc tangent function given in Listing 1.3 can properly handle negative arguments. There are two arguments, corresponding to the x and y components of the angle. This function calls the regular Turbo Pascal ArcTan function with a positive argument. Then the result is corrected for the proper quadrant.

```
FUNCTION Atan(X, Y: Real): Real;
  { arctan in degrees }

CONST
  Pi180 = 57.2957795;

VAR
  A: Real;

BEGIN { Function Atan }
  IF X = 0.0 THEN
    IF Y = 0.0 THEN Atan := 0.0
    ELSE Atan := 90.0
  ELSE   { X <> 0 }
    IF Y = 0.0 THEN   Atan := 0.0
    ELSE   { X and Y <> 0 }
      BEGIN
        A := ArcTan(Abs(Y / X)) * Pi180;
        IF X > 0.0 THEN
          IF Y > 0.0 THEN Atan := A   { X, Y > 0 }
          ELSE  Atan := -A { X>0, Y<0 }
        ELSE      { X < 0 }
          IF Y > 0.0 THEN Atan := 180.0 - A   { X<0, Y>0 }
          ELSE  Atan := 180.0 + A   { X, Y < 0 }
      END { ELSE }
END; { Function Atan }
```

**Listing 1.3:** An arc tangent function with two arguments.

For example, consider an angle which has an x component of $-5$ and a y component of $-5$. This corresponds to an angle of $180° + 45°$ in the third quadrant. However, the tangent of this angle is unity; therefore, the regular arc tangent function returns a value in radians corresponding to the angle 45°. By contrast, the Atan function given in Listing 1.3 properly returns a result of 225° for this example. Notice that this program gives the angle in degrees rather than in the usual radians.

Create a file named ATAN.PAS, and type the routine given in Listing 1.3. We will use this function later in this chapter.

## The Arc Sine and Arc Cosine Functions

The arc sine and arc cosine functions are not included in the Turbo Pascal compilers. However, we can develop our own using the arc tangent function given in Listing 1.3. The Turbo Pascal function shown in Listing 1.4 can calculate the arc sine. It returns an angle in the first quadrant if the argument is positive. If the argument is negative, the program returns a negative angle in the fourth quadrant. The resulting angle is expressed in degrees. Type this routine into a file named ASIN.PAS; we will use it later in this chapter.

The arc cosine function is shown in Listing 1.5. It also requires the arc tangent function given in Listing 1.3. It returns an angle in the first quadrant if the argument is positive. If the argument is negative, the resulting angle is in the second quadrant. We will not use this routine in this book, so you do not need to enter it until you want to perform arc cosine calculations.

```
FUNCTION ArcSin(X: Real): Real;
  { arcsine in degrees }
  { Function Atan is required }

BEGIN { Function ArcSin }
  IF X = 0.0 THEN   ArcSin := 0.0
  ELSE
    IF X = 1.0 THEN ArcSin := 90.0
    ELSE
      IF X = -1.0 THEN ArcSin := -90.0
      ELSE ArcSin := Atan( 1.0, X/ Sqrt(1.0 - Sqr(X)))
END; { Function ArcSin }
```

**Listing 1.4:** An arc sine function.

## Testing the Arc Sine and Arc Tangent Functions

A program that can be used to test the arc functions is given in Listing 1.6. It uses the arc tangent and arc sine functions presented in Listings 1.3 and 1.4. Create a file named TASIN.PAS, and type this program. However, before you run it, let's consider the subject of external files.

```
FUNCTION ArcCos(X: Real): Real;
{ arccosine in degrees }
{ Function Atan is required }

BEGIN { Function ArcCos }
  IF X = 0.0 THEN   ArcCos := 90.0
  ELSE
    IF X = 1.0 THEN ArcCos := 0.0
    ELSE
      IF X = -1.0 THEN ArcCos := 180.0
      ELSE ArcCos := Atan( X/ Sqrt(1.0 - Sqr(X)),1.0)
END; { Function ArcCos }
```

**Listing 1.5:** The arc cosine function.

```
PROGRAM Tasin;
{ test ArcSin and ArcTan }

VAR
  X, Y, Z, Pid180: Real;
  I: Integer;

{$I ATAN.PAS} { Listing 1.3 }
{$I ASIN.PAS} { Listing 1.4 }

BEGIN { main program }
  Pid180 := Pi / 180.0;
  WriteLn('    Sine    Arcsine    Sine');
  FOR I:=1 TO 6 DO
    BEGIN
        X := (I-1) * 0.2;
        Y := ArcSin(X);
        Z := Y * Pid180;
        WriteLn(X:9:4, Y:9:4, Sin(Z):9:4)
    END
END.
```

**Listing 1.6:** Testing the arc sine and arc tangent functions.

## *Using External Files*

Several of the same key routines will be needed in various programs presented in this book. However, you do not need to enter them into each program. Instead, you can store a single copy of each routine as a separate disk file and have the programs refer to that file. This technique has several advantages:

- The main source program will be smaller, less cluttered, and easier to understand.

- Several different source programs can refer to the same external file, thus reducing the amount of disk space required for the programs.

- A routine in an external file is easier to revise since only one copy must be changed.

Listing 1.6 shows how to refer to an external file. It includes the line

```
{$I ATAN.PAS}
```

at the place where the ATAN.PAS routine is needed. This command is known as a compiler INCLUDE directive because it gives a message to the compiler.

Because the INCLUDE directive is enclosed in braces, the characters Turbo Pascal uses for comments, it does not generate computer code. Instead, it directs the compiler to stop processing the statements in the main program and go to the corresponding source file (ATAN.PAS here). The statements in the referenced file are then processed as if they were embedded in the main program. When the compiler completes the INCLUDE file, it returns to the main program. The resulting code will be the same as if the INCLUDE file were embedded in the main source program.

Turbo Pascal INCLUDE directives cannot be nested. That is, an INCLUDE file cannot itself contain an INCLUDE directive to another disk file. However, as shown in Listing 1.6, more than one INCLUDE directive can be used in a program.

Note that the INCLUDE directive is only a bookkeeping feature. Since the INCLUDE file is an ASCII source program, it takes up the same amount of disk space it would as part of the main program. In fact, the total allocated disk space for the main program and the INCLUDE file may be somewhat greater because there is a minimum block size for each disk file. Furthermore, this method does not reduce processing time because the procedure in the INCLUDE file must be compiled each time that the main program is compiled. Nevertheless, the INCLUDE feature is ideal when one procedure is used by several different programs.

Now, run the program shown in Listing 1.6 and check that the result looks like that shown in Figure 1.4. The sine of the arc sine should give the original value.

```
            Sine   Arcsine    Sine
           0.0000    0.0000   0.0000
           0.2000   11.5370   0.2000
           0.4000   23.5782   0.4000
           0.6000   36.8699   0.6000
           0.8000   53.1301   0.8000
           1.0000   90.0000   1.0000
```

**Figure 1.4:** Testing the arc sine and sine functions.

## Summary

In this chapter, we wrote several programs for testing the Turbo Pascal compilers. We began with a program to test floating-point operations; we ran this program on two different compilers and compared the results. Next, we wrote an arc tangent routine that places the angle in the proper quadrant. We then developed our own arc sine and arc cosine functions. We saw how to use the Turbo Pascal INCLUDE directive to incorporate external files into a main source program. Finally, we checked the arc tangent, arc sine, and sine functions.

## Exercises

**1-1.** Write a program to test the square root function, Sqrt. Read a value from the keyboard, take the square root, and display the result. Square this value with the Sqr function and display it on the same line.

**1-2.** Test the arc tangent function by writing a program to calculate $Atn(Sin(X)/Cos(X))$.

**1-3.** Write a program to read values from the keyboard and take Exp(Ln(X)) and Ln(Exp(X)).

**1-4.** Write a function called Power that has two arguments. The result is the first argument to the power of the second.

# *Statistical Tools*

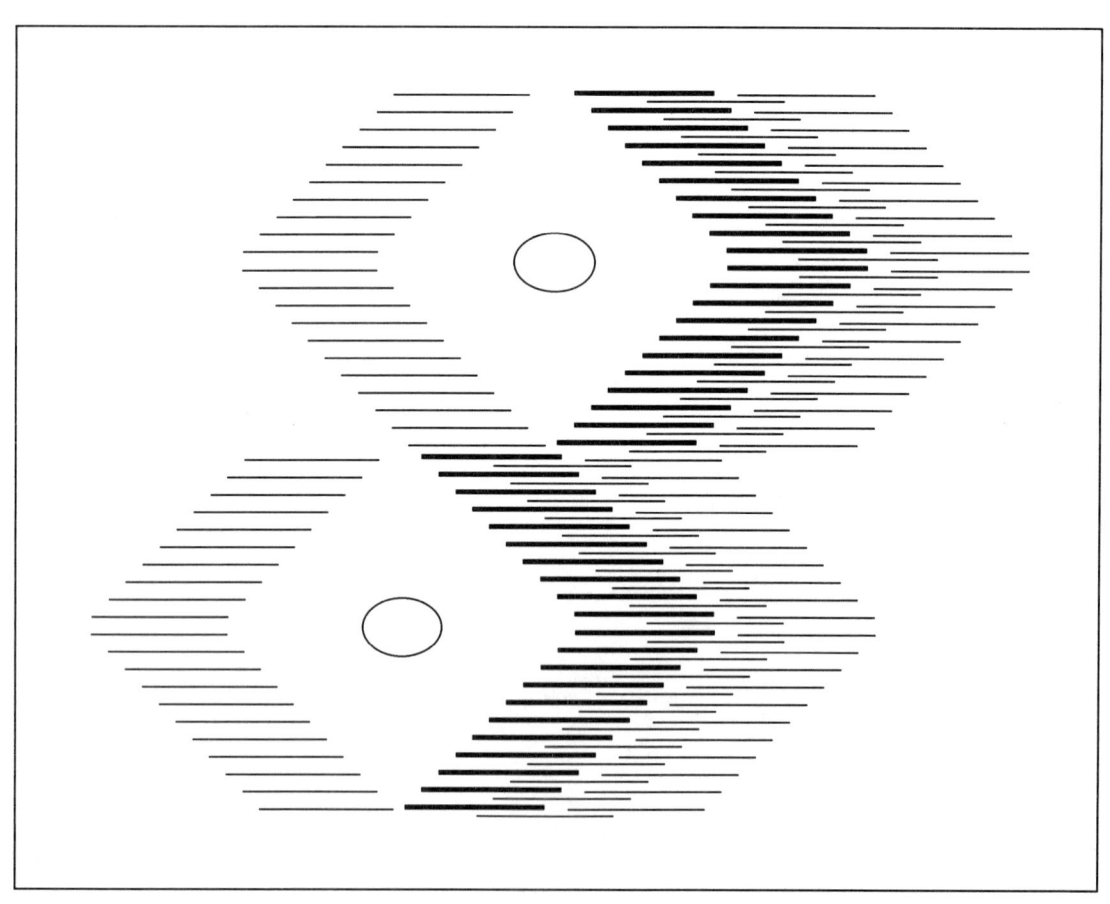

*Chapter* **2**

In this chapter, we will review some statistical tools and discuss how to program them in Turbo Pascal. First, we will describe the uses of the mean and the standard deviation and present a program that calculates both of these values. We will then discuss random numbers and see how to generate them using Turbo Pascal's built-in random-number generator. We will develop another program to evaluate the "randomness" of the numbers produced by this random-number generator. Finally, we will look at a Turbo Pascal function for producing Gaussian-distributed random numbers and test that function.

## The Mean

We often use a single number, called the *average* or *mean value*, to describe a group of data. The mean is calculated by adding all the items in the group and then dividing by the number of items. The formula is

$$\bar{y} = \frac{\sum\limits_{i=1}^{N} y_i}{N} \tag{1}$$

where $y_i$ is the set of data containing $N$ elements. The symbol, $\bar{y}$, (pronounced y-bar) is the resulting mean.

On the other hand, when there is a uniform *continuous distribution* of the data, the mean can be determined by integration. Consider the function $y = f(x)$ over the interval from limit $a$ to limit $b$. The mean value of $y$ is constant over this interval. Therefore, the area under the mean is equal to the area under the curve $f(x)$.

$$\bar{y}(b - a) = \int_a^b f(x)dx$$

Thus, the average value of $y$ is:

$$\bar{y} = \frac{\int_a^b f(x)dx}{b - a}$$

## Dispersion from the Mean

Sometimes all the data are close to the mean. In other cases, there is a great range of values, as with weather averages. For example, the average annual rainfall of San Francisco is reported to be 19 inches, and the average annual snowfall of New York City is given as 30 inches. However, in both cities, some years are very wet, and others are very dry. The average annual temperature of Albuquerque and San Francisco are the same, 57°F, but the climate of these two cities is very different.

As another example, consider a particular brand of breakfast cereal whose box contains the statement

    Net weight 16 ounces

Suppose that an inspector from the Bureau of Weights and Measures decides to check this brand of cereal in a grocery store. Several boxes are opened, and the contents are weighed. The inspector finds that some boxes contain exactly 16 ounces, but others have 15 or 17 ounces. Should the boxes that contain only 15 ounces be confiscated as examples of short weight? If the contents of 100 boxes of cereal were weighed, the resulting frequency distribution might look like the curve shown in Figure 2.1.

The average weight is 16 ounces, but some of the cereal boxes weigh more than the mean value and others weigh less. Furthermore, there are very few boxes that weigh more than 17 ounces or less than 15 ounces.

The frequency distribution shown in Figure 2.1 is *bell-shaped*. The curve shows a *Gaussian*, or *normal*, distribution. This behavior is typical of random variation about a mean value. The equation of this bell-shaped curve is related to the Gamma and Gaussian error functions, which we will study in Chapter 11.

Now suppose that a second type of breakfast cereal is also tested, and the results look like the curve shown in Figure 2.2. The data again show a frequency distribution that is bell-shaped with a mean value of 16 ounces. However, this time, there is a larger dispersion in the weights. Some boxes are as heavy as 26 ounces, while others are as light as 6 ounces.

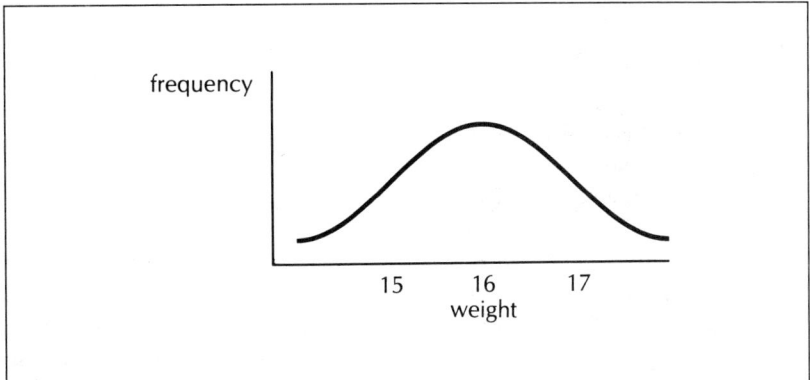

**Figure 2.1:** A bell-shaped curve.

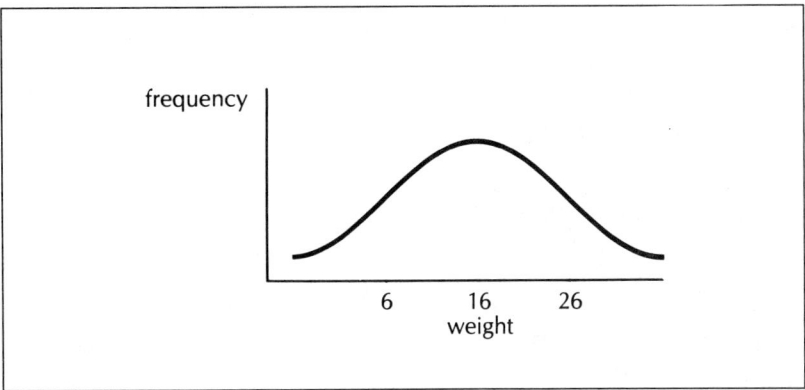

**Figure 2.2:** A bell-shaped curve with a larger dispersion.

Clearly, there is a difference between the packaging of the first type of cereal and the second, even though they both have the same average weight. Something else besides the mean value is needed to describe the distribution. We can characterize the *dispersion* by using the standard deviation, as discussed in the next section.

## The Standard Deviation

The *standard deviation* is a measure of dispersion about the mean value. The symbol for the standard deviation is the lowercase Greek letter sigma ($\sigma$). The standard deviation is defined by the expression

$$\sigma = \sqrt{\frac{\sum\limits_{i=1}^{N} (\bar{y} - y_i)^2}{N - 1}} \tag{2}$$

where $\bar{y}$ is the mean and $y_i$ is the set of $N$ data.

Equation 2 demonstrates the meaning of the standard deviation. The square of the difference between each element and the mean value is important. If the elements are closely grouped about the mean, then this difference will be small, and the corresponding standard deviation will also be small. On the other hand, if data are spread far from the mean, this difference and the resulting standard deviation will both be large.

The standard deviation can be used quantitatively to describe the dispersion of a set of data. For example, a range of 1 standard deviation on either side of the mean includes 68 percent of the sample. About 95 percent of the values lie within 2 standard deviations of the mean, and almost all the values, 99.7 percent, lie within 3 standard deviations of the mean.

$$\bar{y} \pm \sigma \qquad 68\%$$

$$\bar{y} \pm 2\sigma \qquad 95\%$$

$$\bar{y} \pm 3\sigma \qquad 99.7\%$$

Suppose that we want to select a type of steel for building a bridge. We conduct tensile tests and find that one type of steel has a mean strength of 450 megapascals (MPa), with a standard deviation of 10 MPa. The results show that 99.7 percent of the pieces are expected to have a strength in the range of 420 to 480 MPa, that is, within 3 sigmas. Therefore, if we used this steel, our bridge design could be based on a steel strength of 420 MPa, 3 sigmas below the mean value.

Further suppose that we find that a second type of steel has a mean strength of 500 MPa. We want to know if it is a better steel than the first. Our tests show that the standard deviation for this second steel is 40 MPa. This larger sigma means a larger spread in values. A strength of 3 sigmas below the mean of this second steel is only 380 MPa. Thus, this second steel is not as good as the first one, even though its mean strength is higher.

Metals used for construction typically have small sigmas, but brittle materials, such as concrete and glass, usually have large ones. Suppose that we want to find a type of glass suitable for the front door of an office. Our tests on one glass type show that it has an average breaking strength of 120 MPa, with a standard deviation of 40 MPa. In this case, 3 sigmas below the mean gives a value of zero! Clearly, it is important to consider the standard deviation as well as the mean value.

Now that we have discussed the meaning and the importance of the standard deviation, let's see how to calculate this value.

## *Calculation of the Standard Deviation*

Although Equation 2 correctly demonstrates the meaning of the standard deviation, it is an inferior method of calculation. Subtracting each element from the mean value causes a loss of significance. This is especially important for values that are very close to the mean; that is, the smaller the value of sigma, the greater the problem will be. Another disadvantage of Equation 2 is that you must calculate the average before subtracting any values. Thus, you cannot perform this calculation until all the data are available.

A better method of calculating the standard deviation can be derived by expanding the numerator of Equation 2:

$$\Sigma(\bar{y}^2 - 2\bar{y}y_i + y_i^2)$$

Now, by combining the above expression with Equation 1, we have

$$\left(\frac{\Sigma y_i}{N}\right)^2 N - 2\left(\frac{\Sigma y_i}{N}\right)\Sigma y_i + \Sigma y_i^2$$

since $\bar{y} = \Sigma\, y_i/N$. The resulting formula is

$$\sigma = \sqrt{\frac{\Sigma y^2 - \Sigma y\, \Sigma y/N}{N - 1}}$$

This method keeps two running totals: the sum of the individual values and the sum of the squares of the values.

## *Calculating the Mean and Standard Deviation*

Procedure MeanStd, shown in Listing 2.1, is used by the program shown in Listing 2.2 to calculate the mean and standard deviation of a set of numbers. Create a file named MEANSTD, and enter the routine given in Listing 2.1. Create another file named MEANS.PAS, and type the program given in Listing 2.2. Notice that we use an INCLUDE direct-ive in the MEANS program to refer to the routine in the MEANSTD file.

When MEANS is run, it begins by requesting the user to enter a list of items from the keyboard. Then it calls procedure MeanStd to calculate

```
PROCEDURE MeanStd
              ( X: Ary;    { array of values }
                N: Integer;
          VAR Mean: Real;
      VAR Std_Dev: Real);

VAR
   I: Integer;
   Sum_X, Sum_Sq: Real;

BEGIN
   Sum_X := 0;
   Sum_Sq := 0;
   FOR I := 1 TO N DO
     BEGIN
        Sum_X := Sum_X + X[I];
        Sum_Sq := Sum_Sq + X[I] * X[I]
     END;
   Mean := Sum_X / N;
   Std_Dev :=
        Sqrt((Sum_Sq - Sqr( Sum_X) / N)/(N-1))
END; { procedure MeanStd }
```

**Listing 2.1:** Procedure MeanStd to calculate the mean and standard deviation.

```
PROGRAM Means;
{ Find mean and standard deviation }

CONST
  Max = 80;

TYPE
  Ary = ARRAY[1..Max] OF Real;

VAR
  X: Ary;
  I, N: Integer;
  Mean, Std: Real;

{$I MEANSTD.PAS} {Listing 2.1}

BEGIN { main program }
  WriteLn;
  WriteLn(' Calculation of mean and standard deviation');
  REPEAT
    Write(' How many points? ');
    ReadLn(N)
  UNTIL N <= Max;
  FOR I:=1 TO N DO
    BEGIN
      Write( I:3,': ');
      ReadLn(X[I])
    END;
  MeanStd( X, N, Mean, Std);
  WriteLn;
  WriteLn(' For ', N:3, ' points, mean=', Mean:8:4,
          ', sigma=', Std:8:4)
END.
```

**Listing 2.2:** Calculating the mean and standard deviation.

the mean and standard deviation of the list. Procedure MeanStd has four parameters:

- An array of values, declared to be of type Ary
- The number of values in the array
- The mean calculated by the procedure
- The standard deviation calculated by the procedure

The MEANS program shows how arrays can be included as parameters to procedures. Turbo Pascal requires a specific declaration of each parameter. However, the expression

PROCEDURE MeanStd(x: ARRAY[1..max] OF Real; . . .);

is not acceptable. Consequently, each array used as a parameter must be specifically declared in a type statement of the calling program, like this:

```
TYPE
       Ary ARRAY[1..Max] OF Real;
```

Then the defined type is included in the heading of the procedure, like this:

```
PROCEDURE MeanStd(X: Ary; . . .);
```

Our program begins by asking the user for the number of values to be entered. If the answer exceeds the declared length of the array X, then the program repeats the question. After the user enters the values, the program calculates and displays the mean and standard deviation.

If there are many values, you should consider having the computer program count the number of values as they are entered. A special value, such as a number less than -20000, could be used to signal the end of the list. Alternatively, you could have the program read the values from a disk file.

Run the MEANS program, and enter the five numbers 1,2,3,4,5. Verify that the mean is 3.0 and the standard deviation is 1.58. The results should look like Figure 2.3.

This completes our introduction to the mean and the standard deviation. Now we will turn to another statistical tool: random-number generators.

```
Calculation of mean and standard deviation
How many points? 5
  1:  1
  2:  2
  3:  3
  4:  4
  5:  5
For   5 points, mean=  3.0000, sigma=  1.5811
```

**Figure 2.3:** Output from the MEANS program.

# *Random Numbers and Random-Number Generators*

It is sometimes useful to test a computer program with a set of random numbers if the actual data are not available. For example, suppose that we wrote a program to fit a straight line through a set of experimental data. We could generate a set of points along a straight line. Then we could selectively move the points off the line by using a random-number generator (we will do this in Chapter 5).

Although it is not a standard Pascal feature, Turbo Pascal includes a random-number generator. This random-number generator, function Random, will return a sequence of numbers between zero and unity. A typical call to function Random is

```
x[i] := Random;
```

## *Evaluating the Random-Number Generator*

Before you use a random-number generator, you should make a simple test to see how reasonable its results are. The mean value of a set of random numbers ranging from zero to unity should, of course, be one-half. Furthermore, the standard deviation should be the reciprocal of the square root of 12, a value of approximately 0.2887.

The program shown in Listing 2.3 can test a random-number generator. It calls the built-in function Random 48 times. Then it calls procedure MeanStd, given in Listing 2.1, to calculate the mean and standard deviation. This process is repeated 20 times. The program prints the mean value and the corresponding standard deviation for each cycle. It places the expected values at the head of the columns.

Type the program shown in Listing 2.3, and give it the name RANTST. Remember that we have procedure MeanStd in a separate file. Then test the built-in random-number generator. Random-number generators are sometimes called pseudo-random-number generators to emphasize that they are not truly random. Therefore, you should not expect to obtain a mean of one-half and a standard deviation of 0.2887 for each grouping of 48 values. The first part of the output from the RANTST program might look like Figure 2.4.

Notice that the mean values range from 0.40 to 0.56, and the standard deviations go from 0.26 to 0.30. This very good result is characteristic of built-in random-number generators; other random-number generators do not produce values this close to the expected mean and standard deviation.

```
PROGRAM Rantst;
{ test random-number generator Random }
{ procedure MeanStd is required }

TYPE
  Ary = ARRAY[1..100] OF Real;

VAR
  X: Ary;
  N, I, J: Integer;
  Mean, Std: Real;

{$I MEANSTD.PAS } {From Listing 2.1}

BEGIN    { Main program }
  N := 48;
  WriteLn;
  WriteLn( '    mean     std dev');
  WriteLn( '    (0.5)    (0.2887)' );
  WriteLn( '    ====================');
  FOR J := 1 TO 20 DO
    BEGIN
      FOR I := 1 TO N DO
        X[I] := Random;
      MeanStd( X, N, Mean, Std);
      WriteLn( Mean:10:4, Std:10:4)
    END    { J loop }
END.
```

**Listing 2.3:** Testing the built-in random-number generator.

| mean<br>(0.5) | std dev<br>(0.2887) |
|---|---|
| 0.5155 | 0.3024 |
| 0.5059 | 0.2916 |
| 0.5930 | 0.2742 |
| 0.4553 | 0.3073 |
| 0.5104 | 0.2885 |
| 0.5990 | 0.2676 |
| 0.5134 | 0.3227 |
| 0.4982 | 0.2881 |
| 0.5643 | 0.3125 |
| 0.4291 | 0.2625 |
| 0.4738 | 0.2891 |
| 0.4557 | 0.2843 |
| 0.4894 | 0.2554 |
| 0.5236 | 0.3068 |
| 0.4784 | 0.3041 |
| 0.4920 | 0.2775 |
| 0.4305 | 0.2527 |
| 0.5425 | 0.2829 |
| 0.5556 | 0.2741 |
| 0.5625 | 0.2784 |

**Figure 2.4:** Testing the random-number generator with the RANTST program.

## *Using* π *to Produce Random Numbers*

A short sequence of random numbers can be obtained from the first 15 digits of π:

3.14159265358979

The following mnemonic may help you remember the sequence. The number of letters in each word is equal to the corresponding digit of Π:

YES, I NEED A DRINK, ALCOHOLIC, OF COURSE, AFTER
3 .  14   1 5    9        2  6       5

THE HEAVY SESSIONS INVOLVING QUANTUM MECHANICS
3   5    8        9          7        9

This sequence has a mean value of 5.1 and a standard deviation of 2.8.

Our second random-number generator, described in the next section, is designed to simulate experimental data.


## *Gaussian Random Numbers*

Sampling errors, which are errors made during measurement, can be of any size. However, small errors are more likely than large errors. For example, suppose that a 6 foot-long table is measured with a ruler. An error of 1 foot is less likely than an error of 1 inch. An error of 10 feet is even less likely. Thus, measured values that are further from the correct value are less likely than values that are closer. Therefore, if we want to simulate experimental data with a random-number generator, the numbers should not be uniformly spaced as in the previous exercise. However, the Turbo Pascal random-number generator does produce a uniform set of numbers over the interval 0 to 1.

What we need for the simulation of experimental data is a set of random numbers that are grouped about the mean. Thus, the frequency distribution should be bell-shaped; it should have a normal or Gaussian distribution. Fortunately, it is fairly easy to produce a Gaussian distribution of random numbers from an ordinary random-number generator.

Consider a sequence of 12 random numbers. It is highly unlikely that all 12 will have a value of zero. Similarly, it is very unlikely that they will all be unity. Suppose that we generate a new number from the sum of 12 uniformly distributed random numbers. Since the average value of the original numbers is one-half, the sum of 12 random numbers is likely to be about 12 times one-half, or 6. As we will see, this is the key to generating Gaussian random numbers from an ordinary random-number generator. We will now look at a Turbo Pascal version of such a function.

## A Gaussian Random-Number Generator

A Turbo Pascal function for producing Gaussian-distributed random numbers is given in Listing 2.4. Each Gaussian random number is obtained by summing 12 uniformly distributed random numbers and subtracting the value of 6. A set of such numbers should have a mean value of zero and a standard deviation of unity. Other values for the mean and standard deviation can readily be obtained. The formula for calculating each random number, Randg, is

$$\text{Randg} := \text{Sigma} * (\text{Sum} - 6) + \text{Mean}$$

In this expression, Sum is the sum of 12 uniformly distributed random numbers, Sigma is the desired standard deviation, and Mean is the desired mean. If each Gaussian random number is formed from more than 12 numbers, there is an additional complication. In this case, the formula becomes

$$\text{Randg} := \text{Sigma} * (\text{Sum} - \text{Num}/2) * \text{Sqrt}(12/\text{Num}) + \text{Mean}$$

where Num is the number of uniformly distributed random numbers used to obtain each Gaussian random number.

```
FUNCTION Randg(Mean, Sigma: Real): Real;
   { produce random numbers with a Gaussian distribution }
   { Mean and Sigma are supplied by calling program }
   { Function Random is required }

VAR
   I: Integer;
   Sum: Real;

BEGIN
   Sum := 0;
   FOR I := 1 TO 12 DO
     Sum := Sum + Random;
   Randg := (Sum - 6) * Sigma  + Mean
END; { Function Randg }
```

**Listing 2.4:** A function to generate Gaussian distributed random numbers.

Function Randg can thus be used with function Random to generate a series of random numbers with a Gaussian distribution. The calling program must supply two parameters: the desired mean and the standard deviation of the resulting numbers. An alternative approach would be to select fixed values for Mean and Sigma. These could then be encoded into function Randg and would not appear as parameters.

### Evaluating the Gaussian Random-Number Generator

A program for testing function Randg is given in Listing 2.5. The mean for the Gaussian numbers is chosen to be 10, and the standard deviation is chosen to be one-half. Procedure MeanStd is called for the calculation of the mean and standard deviation for a set of the new random numbers. The results, given in Figure 2.5, show that the numbers exhibit a mean of 10 and a standard deviation of one-half.

```
PROGRAM Rantst;
  { test Gaussian random-number generator Randg }
  { procedure MeanStd is required }

TYPE
  Ary = ARRAY[1..100] OF Real;

VAR
  X: Ary;
  N, I, J: Integer;
  Aver, Std: Real;

{$I MEANSTD.PAS } {From Listing 2.1}
{$I RANDG.PAS } {From Listing 2.4}

BEGIN    { Main program }
  N := 50;
  WriteLn;
  WriteLn( '    mean      std dev');
  WriteLn( '    (10)      (0.5)' );
  WriteLn( '    ====================');
  FOR J := 1 TO 20 DO
    BEGIN
      FOR I := 1 TO N DO
        X[I] := Randg(10, 0.5);
      MeanStd( X, N, Aver, Std);
      WriteLn( Aver:10:4, Std:10:4)
    END   { J loop }
END.
```

**Listing 2.5:** Testing the Gaussian random-number generator.

```
            mean      std dev
            (10)      (0.5)
           =====================
            9.9376     0.4948
            9.9724     0.5158
            9.9396     0.5354
            9.9978     0.5172
           10.0609     0.4827
           10.0455     0.5313
           10.1183     0.4574
            9.8253     0.5123
            9.9223     0.4686
           10.0928     0.5737
            9.9762     0.4993
           10.0028     0.4678
           10.0328     0.5261
            9.9923     0.4725
            9.9623     0.5064
           10.1397     0.4790
            9.9275     0.4499
           10.0479     0.5189
           10.0981     0.5378
           10.0036     0.5298
```

**Figure 2.5:** Gaussian random numbers.

The area under the normal distribution curve is related to the standard deviation. It can be obtained from the Gaussian error function that we will develop in Chapter 11.

## Summary

We began this chapter with a discussion of two important statistical tools: the mean and the standard deviation. Our discussion led us to an efficient Turbo Pascal program for calculating these values. Then we discovered how to generate two different types of random numbers. We discussed the importance of evaluating the random numbers generated, and we saw two programs that can perform such an evaluation.

In the next chapter, we will continue to expand our understanding of Turbo Pascal; we will see how to express vectors and matrices as arrays.

## Exercises

**2-1.** Alter the program given in Listing 2.2 so that the number of items to be averaged is determined by the program. Use a value less than -20000 to signal the end of data. The program should repeatedly loop while reading items from the keyboard until a number less than -20000 is entered.

**2-2.** Alter the program given in Listing 2.2 so that the items to be averaged are obtained from procedure Get_Data rather than from the keyboard. Replace the ReadLn statement in the main program with a call to this procedure. Use the value $-20000$ as a signal for the end of data. The program should repeatedly loop, receiving values and incrementing the number of items, $N$, until the number -20000 is encountered.

**2-3.** Verify that the first 15 digits of $\Pi$ have an average value of 5.1 and a standard deviation of 2.8.

**2-4.** Alter the function given in Listing 2.4 so that 24 uniformly distributed random numbers are used to generate each Gaussian random number. Run the complete program and compare the results with the original.

# Vector and Matrix Operations

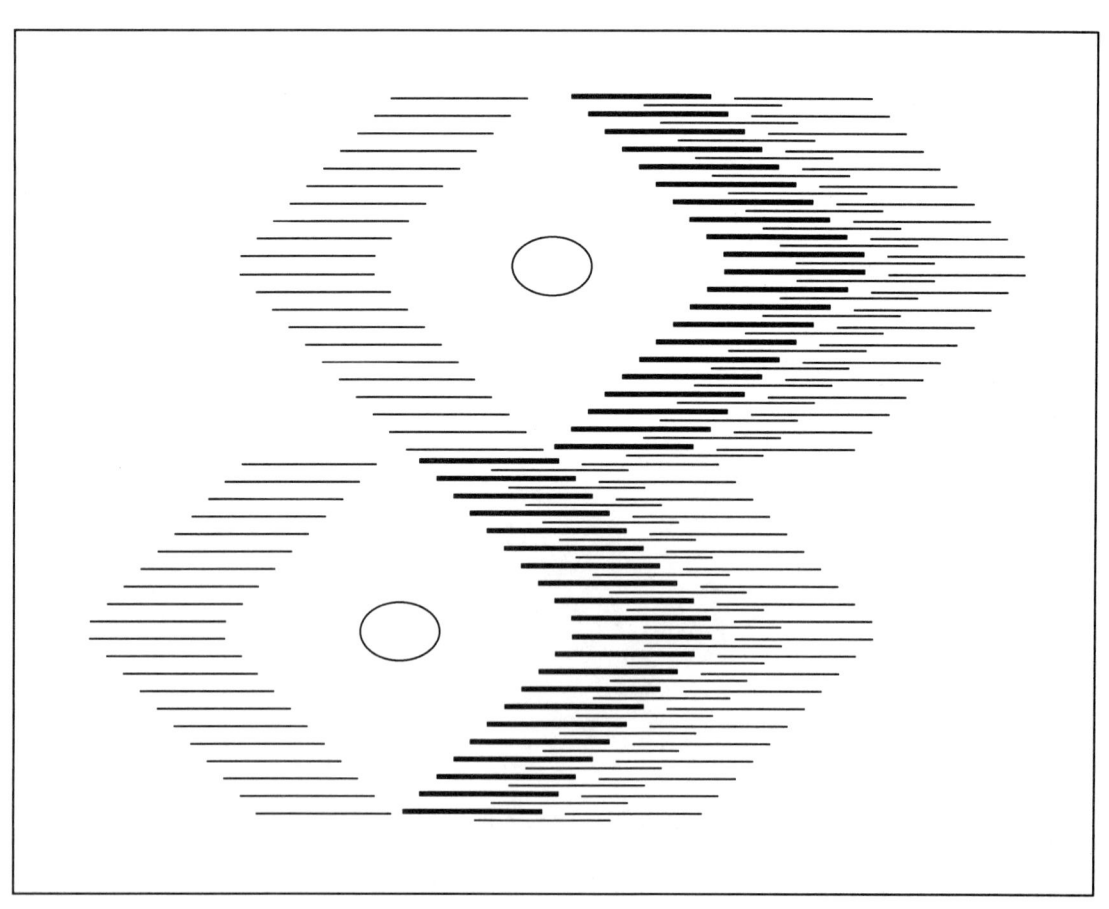

Vector and matrix structures are useful for calculating mathematical operations on data sets, and we will develop programs that use them throughout this book. In this chapter, we will review vector and matrix mathematical operations and consider how to program them in Turbo Pascal.

## *Scalars and Arrays*

In Turbo Pascal programs, vectors and matrices are represented as arrays, which are sets of *scalars*. An ordinary, simple variable is called a scalar. It is referenced by a unique, symbolic name, and it is associated with a single value. For example, the Turbo Pascal expression

```
Terms := 12
```

assigns a value to the scalar variable Terms. In Turbo Pascal, scalars can be declared as real, integer, character, or logical (Boolean) variables.

Sometimes we may need to reference a set of scalar values collectively. Turbo Pascal provides several structured data types for this purpose, but the most useful of these is the *array*. In an array, all the components have the same type. That is, all the elements are real numbers, integers, characters, or logical variables. Turbo Pascal allows structures within structures, which means that the elements of an array can themselves be arrays. However, for the programs in this book, the elements of an array are scalars.

The components of an array are collectively referenced by a symbolic name. Each element of the array is uniquely identified by an index, or subscript, that follows the name. This index allows us to access each value in the array individually. We can change one value in the array without affecting the other values.

In the following sections, we will see how vectors and matrices are treated as arrays in Turbo Pascal programs.

## *Vectors*

A vector is a one-dimensional array. The number of *dimensions* refers to the number of subscripts, not the number of elements. Each element of a vector is referenced through an index that usually runs from 1 up to the maximum number of elements of the vector. However, in Turbo Pascal, the index can begin at any point. Thus, an array index might begin with 0, or it might begin with 1978.

In ordinary usage, the elements of a vector are separated by spaces. For example, consider the vector V that contains the values

    2 5 1 9 4 3

The Turbo Pascal statement

    VAR V: ARRAY[1..6] OF Integer;

defines the symbol V as a vector of integers. The maximum number of elements (the length) is declared to be six. We can assign the above values to this vector by using the Turbo Pascal statements

    V[1] := 2;
    V[2] := 5;
    V[3] := 1;
    V[4] := 9;
    V[5] := 4;
    V[6] := 3;

The first element of this vector is located at position V[1], the second element is at V[2], and the last element is at V[6]. The first element has a value of 2, the second element has a value of 5, and the last element has a value of 3.

Since vectors have only one dimension, it does not usually matter whether they are written horizontally or vertically. However, when we multiply vectors with matrices, we need to distinguish between *row vectors*, which are written horizontally, and *column vectors*, which are written vertically. The set

$$\begin{bmatrix} 2 \\ 5 \\ 1 \\ 9 \\ 4 \\ 3 \end{bmatrix}$$

is a column vector.

In the next section, we will study the operations of vector arithmetic and see how to program vectors in Turbo Pascal.

## Vector Arithmetic

The principal arithmetic operations defined for vectors are magnitude, scalar multiplication, vector addition, dot product, and cross product. Let's look at each of these operations in detail.

### Magnitude

The *magnitude* of a vector is a scalar value. It is obtained by summing the squares of the elements, then taking the square root of the resulting sum. For example, the magnitude of the vector Y:

$$Y = [2 \quad 2 \quad 1]$$

is equal to the square root of $4 + 4 + 1$. The result is 3. We can program the operation by using the Turbo Pascal statement

```
Mag := Sqrt(Sqr(Y[1]) + Sqr(Y[2]) + Sqr(Y[3]))
```

### Scalar Multiplication of Vectors

If a vector V is multiplied by a scalar value S, then each element of V is multiplied by S. For example, two times the vector V is

$$2V = [4 \quad 10 \quad 2 \quad 18 \quad 8 \quad 6]$$

The following Turbo Pascal expression generates a vector V2, in which each element is twice as large as the corresponding element of vector V:

```
FOR I := 1 TO 6 DO
    V2[I] := 2.0 * V[I]
```

### Vector Addition

Two vectors can be added if each has the same number of elements, or the same length. The result is a new vector in which each element is the sum of the two corresponding elements of the original vectors. Thus, if

$$A = \begin{bmatrix} 1 & 2 & 3 \end{bmatrix}$$

and

$$B = \begin{bmatrix} 3 & 4 & 5 \end{bmatrix}$$

then

$$A + B = \begin{bmatrix} 4 & 6 & 8 \end{bmatrix}$$

The corresponding Turbo Pascal expression is

```
FOR I := 1 TO 3 DO
      C[I] := A[I] + B[I]
```

### Dot (or Scalar) Product

The *dot*, or *scalar, product* of two equal-length vectors produces a scalar result. Each element of one vector is multiplied by the corresponding element of the other. The resulting products are then added together. The mathematical symbol for the dot operator is simply a dot placed between the operands, as in the following example:

$$A \cdot B = (1)(3) + (2)(4) + (3)(5) = 26$$

The dot product is equal to the product of the magnitude of the original vectors and the cosine of the angle between them:

$$A \cdot B = |A| \, |B| \, \text{Cos } \theta$$

This formula can be used to find the angle between two vectors. For example, we would calculate the angle between the vectors A and B as

$$\text{Cos } \theta = 26/(3.742 \cdot 7.071) = 0.9826 \quad \text{and}$$

$$\text{ArcCos } 0.9826 = 10.7°$$

### Cross (or Vector) Product

The *cross*, or *vector, product* of two vectors produces a third vector that is mutually perpendicular to the two original vectors. The mathematical symbol for this operation is $\times$. The magnitude of the resulting vector is equal to the product of the magnitudes of the original vectors and the sine of the angle between them:

$$|A \times B| = |A| \, |B| \, \text{Sin } \theta$$

The cross product, $C = A \times B$, can be calculated with the Turbo Pascal expressions

```
C[1] := A[2]*B[3] - B[2]*A[3];
C[2] := -A[1]*B[3] + B[1]*A[3];
C[3] := A[1]*B[2] - B[1]*A[2]
```

The cross product of the vectors $A$ and $B$ is the vector $[-2 \;\; 4 \;\; -2]$; its magnitude is 4.9.

We have now seen Turbo Pascal implementations for the main arithmetic operations on vectors. Later in this chapter, we will see the matrix equivalents of these operations. However, before moving on to matrices, we will define one special type of vector: the string.


## *Strings*

*Strings* of alphabetic and numeric (alphanumeric) characters are useful for representing items such as names and addresses. With standard Pascal, you must define a string as an array of characters and fix its length by using a DECLARATION statement. Each time that you use such a string, it must contain exactly the number of declared characters. However, this is not necessary with Turbo Pascal because it includes the string type, which greatly simplifies string operations.

You define a string variable in Turbo Pascal with a TYPE statement that specifies its maximum length. The string can contain less than the number of characters specified. For example, the Turbo Pascal expression

```
TYPE
    Address = String[15];
```

declares the symbol Address to be a string with a maximum number of 15 characters. Then the statement

```
VAR
    Mail_List: ARRAY [1..100] OF Address;
```

establishes the symbol Mail_List as an array of strings that can contain up to 15 characters each.

Sorting and other operations can be done on strings. For example, the expression

    IF Mail_List[I] < Mail_List[J] THEN . . .

will compare the Ith string to the Jth string. You can also establish a string constant for easy reference. For example, the statement

    CONST
            Heading = ' Least Squares Fit'

allows you to display the string with the statement

    WriteLn(Heading)

Now that we have described vectors, let's move on to two-dimensional arrays and their Turbo Pascal representations.

## *Matrices*

A two-dimensional array is called a *matrix*. The elements of this set are arranged in a square or rectangle. These elements can be considered as a set of horizontal lines called row vectors or as a set of vertical lines called column vectors. Thus, a matrix can be described as a one-dimensional set of vectors.

Each element of a matrix is uniquely defined by a pair of indices: the *row index* and the *column index*. For example, consider the matrix

$$\begin{bmatrix} x_{11} & x_{12} & x_{13} & \cdots & x_{1m} \\ x_{21} & x_{22} & x_{23} & \cdots & x_{2m} \\ x_{31} & x_{32} & x_{33} & \cdots & x_{3m} \\ \cdots & \cdots & \cdots & \cdots & \cdots \\ x_{n1} & x_{n2} & x_{n3} & \cdots & x_{nm} \end{bmatrix}$$

which contains $N$ rows and $M$ columns. The row index is always given before the column index.

A matrix is referenced by its name, which can be a single alphabetic character or a string of characters. The indices appear as subscripts, except in computer programs. In COBOL, FORTRAN, and BASIC programs, the array subscripts are enclosed in parentheses. Square brackets are used for this purpose in Turbo Pascal and APL. Thus, the expression X[2,3] in a Turbo Pascal program refers to the element located in the second row and the third column of the two-dimensional array named X.

A matrix that has the same number of rows as columns is called a *square matrix*. The *principal*, or *main*, *diagonal* of a square matrix contains the elements $X_{11}$, $X_{22}$, $X_{33}$, . . . $X_{nn}$. The term principal diagonal is sometimes shortened to simply the *diagonal*. A square matrix that contains the value of unity at each position of the main diagonal, and is zero everywhere else, is known as a *unit matrix*. For example:

$$\begin{bmatrix} 1 & 0 & 0 & 0 \\ 0 & 1 & 0 & 0 \\ 0 & 0 & 1 & 0 \\ 0 & 0 & 0 & 1 \end{bmatrix}$$

is a 4-by-4 unit matrix. We will consider the unit matrix again later in this chapter and in the next chapter.

Now that we have defined the basic vocabulary of matrices, let's explore the major arithmetic operations for them and the Turbo Pascal implementations of these operations.

## Matrix Arithmetic

The arithmetic operations defined for matrices are transpose; scalar multiplication; and matrix addition, subtraction, and multiplication. We'll discuss each in detail.

### The Transpose Operation

A matrix is *transposed* by interchanging the rows and the columns. Each original element $X_{ij}$ becomes the new element $X_{ji}$ of the transposed matrix. The transpose of matrix X is designated as $X^T$. Thus, if

$$X = \begin{bmatrix} 1 & 2 & 3 \\ 4 & 5 & 6 \\ 7 & 8 & 9 \end{bmatrix}$$

then

$$X^T = \begin{bmatrix} 1 & 4 & 7 \\ 2 & 5 & 8 \\ 3 & 6 & 9 \end{bmatrix}$$

Note that the transpose of a square matrix can be obtained by rotating the matrix about the principal diagonal.

Two matrices are *equal* if every element of one is equal to the corresponding element of the other. Thus, if $X = Y$, then $X_{ij} = Y_{ij}$ for each element. A square matrix is *symmetric* if it is equal to its transpose. Then each element $X_{ij}$ equals the corresponding element $X_{ji}$.

### Scalar Multiplication of Matrices

If a matrix $X$ is multiplied by a scalar value S, then each element of the matrix is multiplied by the value S. The following Turbo Pascal statement will produce a matrix $Y$ from the product of matrix $X$ and the scalar S:

```
FOR I := 1 TO N DO
   FOR J := 1 TO M DO
      Y[I,J] := X[I,J] * S
```

### Matrix Addition and Subtraction

One matrix can be added to or subtracted from another matrix if they both have the same number of rows and columns. The addition of matrix $X$ to matrix $Y$ to produce matrix $Z$ is written as $Z = X + Y$.
   Thus, if

$$X = \begin{bmatrix} x_{11} & x_{12} & x_{13} \\ x_{21} & x_{22} & x_{23} \end{bmatrix}$$

and

$$Y = \begin{bmatrix} y_{11} & y_{12} & y_{13} \\ y_{21} & y_{22} & y_{23} \end{bmatrix}$$

then

$$Z = \begin{bmatrix} x_{11} + y_{11} & x_{12} + y_{12} & x_{13} + y_{13} \\ x_{21} + y_{21} & x_{22} + y_{22} & x_{23} + y_{23} \end{bmatrix}$$

The corresponding Turbo Pascal statement is

```
FOR I := 1 TO N DO
   FOR J := 1 TO M DO
      Z[I,J] := X[I,J] + Y[I,J]
```

In this statement, each element of $Z$ is formed from the sum of the corresponding elements of $X$ and $Y$. In a similar way, we can subtract one matrix from another, as in $Z = X - Y$, by subtracting each element of the second matrix from the corresponding element of the first.

### Matrix Multiplication

One matrix can be multiplied by another if the number of columns of the first matrix equals the number of rows of the second. Then the two matrices are said to be *conformable*. Thus, if X is a matrix that contains $M$ rows and $N$ columns, and Y is a matrix with $N$ rows and $P$ columns, then $Z = XY$ produces a matrix Z with $M$ rows and $P$ columns. That is, the resulting matrix has the same number of rows as matrix X and the same number of columns as matrix Y.

In matrix multiplication, each element of matrix Z is formed from a sum of products. The elements from one row of matrix X are each multiplied by the corresponding elements from a column of matrix Y, then the terms are totaled. For matrix X (which might be the transpose of the previous X)

$$X = \begin{bmatrix} x_{11} & x_{12} \\ x_{21} & x_{22} \\ x_{31} & x_{32} \end{bmatrix}$$

and matrix Y

$$Y = \begin{bmatrix} y_{11} & y_{12} & y_{13} \\ y_{21} & y_{22} & y_{23} \end{bmatrix}$$

the operation $Z = XY$ is formed as

$$Z = \begin{bmatrix} x_{11}y_{11} + x_{12}y_{21} & x_{11}y_{12} + x_{12}y_{22} & x_{11}y_{13} + x_{12}y_{23} \\ x_{21}y_{11} + x_{22}y_{21} & x_{21}y_{12} + x_{22}y_{22} & x_{21}y_{13} + x_{22}y_{23} \\ x_{31}y_{11} + x_{32}y_{21} & x_{31}y_{12} + x_{32}y_{22} & x_{31}y_{13} + x_{32}y_{23} \end{bmatrix}$$

Each $jk$ element of matrix Z is formed from row $j$ of matrix X and column $k$ of matrix Y according to the scheme

$$z_{jk} = x_{j1}y_{1k} + x_{j2}y_{2k} + \cdots + x_{jn}y_{nk}$$

Thus if

$$X = \begin{bmatrix} 1 & 4 \\ 2 & 5 \\ 3 & 6 \end{bmatrix}$$

and

$$Y = \begin{bmatrix} 7 & 8 & 9 \\ 10 & 11 & 12 \end{bmatrix}$$

then the product

$$Z = XY = \begin{bmatrix} 47 & 52 & 57 \\ 64 & 71 & 78 \\ 81 & 90 & 99 \end{bmatrix}$$

is a square matrix with dimensions of 3 by 3. The first element, $Z_{11}$, for example, is calculated as

$$(1)(7) + (4)(10) = 47$$

Matrix multiplication is not commutative; that is, the product YX will not, in general, be equal to the product XY. Reversing the order of the previous example produces a 2-by-2 matrix, rather than a 3-by-3 matrix

$$YX = \begin{bmatrix} 50 & 122 \\ 68 & 167 \end{bmatrix}$$

Notice that the dot product of two vectors follows the rules for matrix multiplication if the first vector is a column vector (that is, has one column) and the second is a row vector (that is, has one row).

We have seen how Turbo Pascal handles vector and matrix arithmetic with one- and two-dimensional arrays. We are now ready to write a Turbo Pascal program using the statements reviewed in the preceding sections.

## *Programming Matrix Multiplication*

In Chapter 4, when we deal with linear equations, we will need a routine for matrix multiplication. Specifically, we will need to multiply the transpose of a matrix X by the original matrix X. We will also need to multiply the vector Y by the matrix X. Let's go ahead and program this routine now.

The program shown in Listing 3.1 will do both of the multiplication operations. The main program will generate the X matrix and the Y vector. Procedure Square, given in Listing 3.2, will calculate the needed matrix A and vector G according to the equations

$$X^T X = A \quad \text{and} \quad YX = G$$

```
PROGRAM Matr1;
{ Turbo Pascal program to perform }
{ matrix multiplication }

CONST
  Rmax = 9;
  Cmax = 3;

TYPE
  Ary   = ARRAY[1..Rmax] OF Real;
  Arys  = ARRAY[1..Cmax] OF Real;
  Ary2  = ARRAY[1..Rmax, 1..Cmax] OF Real;
  Ary2s = ARRAY[1..Cmax, 1..Cmax] OF Real;

VAR
  Y: Ary;
  G: Arys;
  X: Ary2;
  A: Ary2s;
  Nrow, Ncol: Integer;

PROCEDURE Get_Data(VAR X: Ary2;
                   VAR Y: Ary;
                VAR Nrow, Ncol: Integer);

{ get values for Nrow, Ncol, and arrays X, Y }

VAR
  I, J: Integer;

BEGIN
  Nrow := 5;
  Ncol := 3;
  FOR I := 1 TO Nrow DO
    BEGIN
      X[I,1] := 1;
      FOR J := 2 TO Ncol DO
        X[I,J] := I * X[I,J-1];
      Y[I] := 2 * I
    END
END; { procedure Get_Data }

PROCEDURE Write_data;

{ print out the answers }

VAR
  I, J: Integer;
```

**Listing 3.1:** Matrix multiplication program (A = $X^TX$, G = YX).

```
  BEGIN
    WriteLn;
    WriteLn('              X                    Y');
    FOR I := 1 TO Nrow DO
      BEGIN
        FOR J := 1 TO Ncol DO
          Write( X[I,J]:7:1, '  ');
        WriteLn( ' : ', Y[I]:7:1)
      END;
    WriteLn('              A                    G');
    FOR I := 1 TO Ncol DO
      BEGIN
        FOR J := 1 TO Ncol DO
          Write( A[I,J]:7:1, '  ');
        WriteLn( ' : ', G[I]:7:1)
      END
  END; { Write_data }

  {$I SQUARE.PAS} {Listing 3.2}

  BEGIN  { main program }
    Get_Data (X, Y, Nrow, Ncol);
    Square(X, Y, A, G, Nrow, Ncol);
    Write_data
  END.
```

**Listing 3.1:** Matrix multiplication program ($A = X^TX$, $G = YX$) (continued).

The matrix X, in this example, contains 5 rows and 3 columns. Vector Y has a length of 5.

Since the resulting matrix A is symmetric, the calculations can be simplified. For example, there will be terms like

$$A[1,3] := A[3,1]$$

Matrix A will have 3 rows (the same as the transpose of matrix X) and 3 columns (the same as matrix X). Vector G will have a length of 3. Actually, both Y and G must be considered as row vectors, that is, they have a dimension of 1 row and 5 columns. Alternatively, if we want to consider the vectors Y and G as column vectors, then we should write the multiplication equation as

$$Y^TX = G^T$$

where the transposed column vectors become row vectors.

Create the procedure given in Listing 3.2, and save it with the name SQUARE.PAS. We will use this procedure in later chapters. Enter the program shown in Listing 3.1, and give it the name MATR1.PAS. Notice that the INCLUDE directive refers to procedure Square. Run the program and check the results with those shown in Figure 3.1.

```
PROCEDURE Square(X: Ary2;
                 Y: Ary;
             VAR A: Ary2s;
             VAR G: Arys;
         Nrow,Ncol: Integer);

{ matrix multiplication routine }
{ A = transpose X times X }
{ G = Y times X }

VAR
  I, K, L: Integer;

BEGIN  { Square }
  FOR K := 1 TO Ncol DO
    BEGIN
      FOR L := 1 TO K DO
        BEGIN
          A[K,L] := 0;
          FOR I := 1 TO Nrow DO
            BEGIN
              A[K,L] := A[K,L] + X[I,L] * X[I,K];
              IF K <> L THEN A[L,K] := A[K,L]
            END
        END { L loop };
      G[K] := 0;
      FOR I := 1 TO Nrow DO
        G[K] := G[K] + Y[I] * X[I,K]
    END  { K loop }
END; { Square }
```

**Listing 3.2:** Procedure Square.

```
           X                              Y
   1.0      1.0       1.0       :     2.0
   1.0      2.0       4.0       :     4.0
   1.0      3.0       9.0       :     6.0
   1.0      4.0      16.0       :     8.0
   1.0      5.0      25.0       :    10.0
           A                              G
   5.0     15.0      55.0       :    30.0
  15.0     55.0     225.0       :   110.0
  55.0    225.0     979.0       :   450.0
```

**Figure 3.1:** Output from the matrix multiplication program.

## Determinants

The *determinant* of a square matrix X is designated as $|X|$. The result is a scalar value. For a 2-by-2 matrix, the upper-left member is multiplied by the lower-right, and then the product of the lower-left member and the upper-right member is subtracted:

$$|X| = x_{11}x_{22} - x_{12}x_{21}$$

For example, the determinant of the matrix

$$\begin{bmatrix} 1 & 2 \\ 3 & 4 \end{bmatrix}$$

is $-2$.

You can calculate the determinant of matrices larger than 2 by 2 by multiplying each element of the first row by the determinant of the remaining matrix, after removing the column that is common to the element in the first row. A recursive definition is

$$|X| = x_{11}s_{11} - x_{12}s_{12} + x_{13}s_{13} - \cdots (-1)^{n+1}x_{1n}s_{1n}$$

where $X_{11}$, $X_{12}$, etc., are the elements of the first row of matrix X, and $S_{1n}$ is the determinant of the matrix that has row 1 and column $n$ removed. The determinant of the matrix

$$\begin{bmatrix} 1 & 2 & 3 \\ 4 & 5 & 6 \\ 7 & 8 & 0 \end{bmatrix}$$

is equal to

$$1\begin{vmatrix} 5 & 6 \\ 8 & 0 \end{vmatrix} - 2\begin{vmatrix} 4 & 6 \\ 7 & 0 \end{vmatrix} + 3\begin{vmatrix} 4 & 5 \\ 7 & 8 \end{vmatrix}$$

The next step is to evaluate the minors

$$\begin{vmatrix} 5 & 6 \\ 8 & 0 \end{vmatrix} \quad \begin{vmatrix} 4 & 6 \\ 7 & 0 \end{vmatrix} \quad \text{and} \quad \begin{vmatrix} 4 & 5 \\ 7 & 8 \end{vmatrix}$$

according to the procedure

$$1(5 \cdot 0 - 8 \cdot 6) - 2(4 \cdot 0 - 7 \cdot 6) + 3(4 \cdot 8 - 7 \cdot 5)$$

The resulting value of the determinant is 27 in this example. If each element of a row or column is zero, then the determinant is zero. Also, if two rows or columns are identical, then the determinant is zero.

We have described the method for calculating the determinant of a matrix. Since we will be using determinants in later chapters, now we will consider a Turbo Pascal program that calculates them.

## *Calculating Determinants*

A program that can be used to find the determinant of a 3-by-3 matrix is given in Listing 3.3. Create a file named DETER.PAS, and

```
PROGRAM Determ;
{ Turbo Pascal program to calculate  }
{ the determinant of a 3-by-3 matrix }

TYPE
  Ary2  = ARRAY[1..3, 1..3] OF Real;

VAR
  A: Ary2;
  N: Integer;
  YesNo: Char;
  D: Real;

PROCEDURE Get_Data(VAR A: Ary2;
                       N: Integer);
{ get values for N and arrays X, Y }

VAR
  I, J: Integer;

BEGIN
  N := 3;
  WriteLn;
  FOR I := 1 TO N DO
    BEGIN
      FOR J := 1 TO N DO
        BEGIN
          Write( J:3, ': ');
          ReadLn( A[I,J])
        END  { J loop }
    END; { I loop }
  WriteLn;
  FOR I:= 1 TO N  DO
    BEGIN
      FOR J:= 1 TO N DO
        Write( A[I,J]:7:4, '  ');
      WriteLn
    END:
```

**Listing 3.3:** The determinant of a 3-by-3 matrix.

```
       WriteLn
END; { procedure Get_Data }

FUNCTION Deter (A: Ary2): Real;
{ calculate the determinant of a 3-by-3 matrix }

VAR
   Sum: Real;

BEGIN
   Sum := A[1,1] * (A[2,2]*A[3,3] - A[3,2]*A[2,3])
        - A[1,2] * (A[2,1]*A[3,3] - A[3,1]*A[2,3])
        + A[1,3] * (A[2,1]*A[3,2] - A[3,1]*A[2,2]);
   Deter := Sum
END;

BEGIN  { main program }
   REPEAT
      Get_Data (A, N);
      D := Deter(A);
      WriteLn( 'The determinant is ', D:12);
      WriteLn;
      Write(' More? ');
      ReadLn( YesNo)
   UNTIL (UpCase(YesNo) <> 'Y')
END.
```

**Listing 3.3:** The determinant of a 3-by-3 matrix (continued).

type the program. When you run this program, procedure Get_Data asks the user to enter the matrix elements. Function Deter then calculates the determinant. The elements of the matrix are entered row by row; that is, the order is

A[1,1], A[1,2], A[1,3], A[2,1], A[2,2],...

After you have entered the nine elements, the program displays the value of the determinant. Then the question "More?" appears. Respond with a Y if you want to calculate the determinant of another matrix.

## Inverse Matrices and Matrix Division

The *inverse* of a nonsingular matrix $X$ is written as $X^{-1}$. The inverse of a singular matrix is undefined. The product of a matrix and its inverse is the identity matrix $XX^{-1} = I$. Matrix inversion is similar to other inverse operations, since the inverse of an inverse produces the original matrix $(X^{-1})^{-1} = X$. If this calculation is done with a computer, however, the two will not always agree because of roundoff error.

*Matrix division* is calculated by a combination of inversion and multiplication. Thus, if we need to divide matrix X by matrix Y, we first invert matrix Y and then multiply the result by matrix X. The operation is written as $XY^{-1}$.

We are interested in matrix inversion because it can give us the solution to a set of simultaneous linear equations. Thus, if we have a coefficient matrix A and a constant vector Y, we want a solution vector B such that $AB = Y$. Then the solution can be obtained from a product of the inverse of matrix A and the constant vector Y (in that order), $AY^{-1} = B$. We will develop matrix-inversion routines in the next chapter.

## Summary

In this chapter, we saw how vectors and matrices are represented in Turbo Pascal. We also studied the methods of programming matrix and vector arithmetic operations and developed two significant programs—one for matrix multiplication and one for calculating determinants. We will use these programs in the following chapters.

## Exercises

**3-1.** Write a program that will read two (three-component) vectors from the keyboard, calculate the dot product, and then display the answer. Show that the dot product of the vectors

$$a = [1\ 1\ 1] \text{ and } b = [1\ -1\ -1]$$

is $-1$.

**3-2.** Write a program that will read two vectors from the keyboard, calculate the cross product, and then display the answer. Show that the cross product of the vectors

$$a = [1\ 1\ 1] \text{ and } b = [1\ -1\ 0]$$

is the vector $[1\ 1\ -2]$.

**3-3.** Write a program that will read two vectors from the keyboard, calculate the angle between the two vectors, and then display the answer. Show that the angle between the vectors

a = [1 1 1] and b = [1 −1 −1]

is 109.5°. (This is the O-Si-O bond angle in silicate minerals.)

**3-4.** A unit vector has a magnitude of unity. Dividing a vector by its magnitude produces a unit vector. Write a program to convert a three-component vector entered from the keyboard into a unit vector. Run the program to show that the vector

a = [18 −18 9]

has the unit vector [0.66667 −0.66667 0.33333]

# Simultaneous Solution of Linear Equations

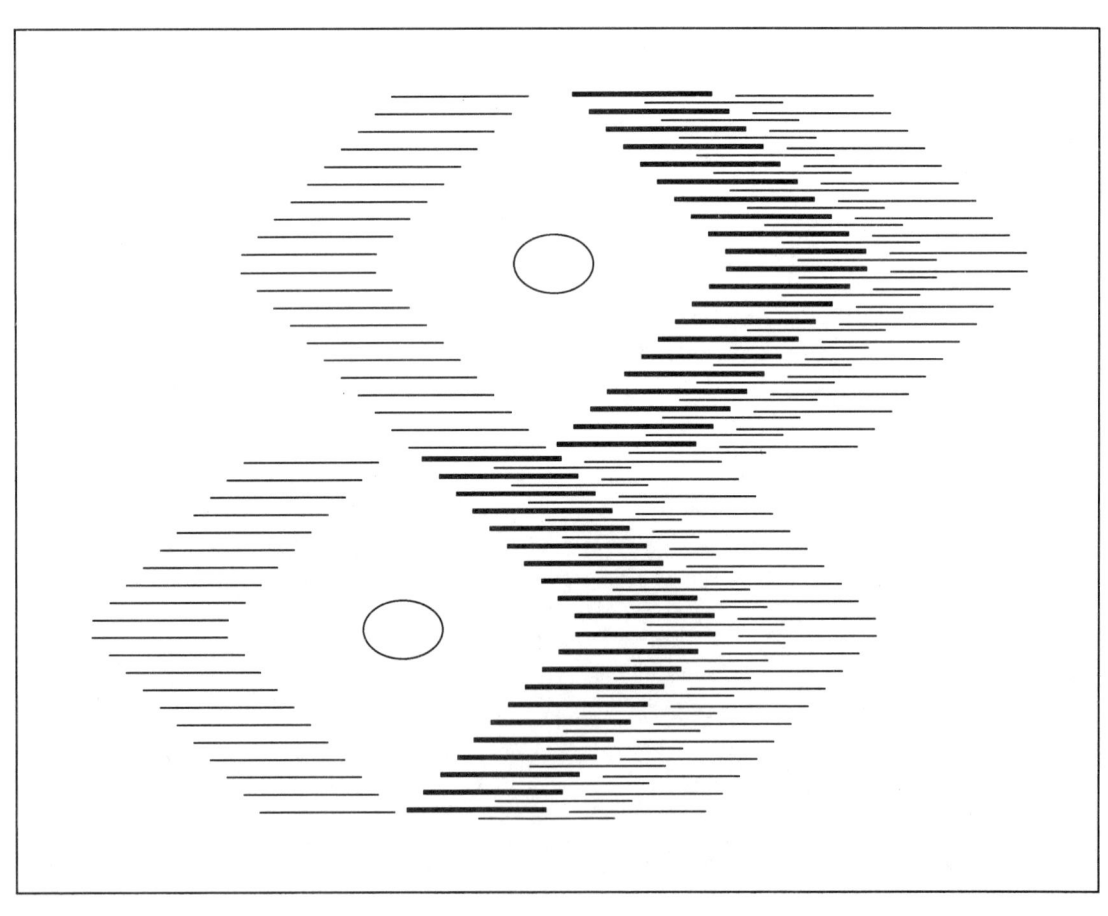

*Chapter* **4**

In this chapter, we will consider the simultaneous solution of a set of linear equations. We will use Cramer's rule, the Gauss elimination method, the Gauss-Jordan elimination method, and the Gauss-Seidel method. In addition, we will study the problem of ill conditioning by generating a set of Hilbert matrices.

We will develop several special Turbo Pascal application programs, including those that solve equations with multiple constant vectors and complex coefficients. Another version will read the data from a disk file instead of from the keyboard. We will also create a program for producing the best fit to an overdetermined system.

We begin by describing linear equations and providing a simple example of simultaneous solution.

# Linear Equations and Simultaneous Solutions

A *linear equation* consists of a sum of terms, such as

$$Ax + By + Cz = D$$

In this example, x, y, and z are variables, and A, B, C, and D are constants. No more than one variable can occur in each term, and the variables must be present to the first power. Thus, expressions such as

$$2x^2 + 3y^2 = 4$$

$$\mathrm{Sin}(x) + \mathrm{Log}(y) = 9$$

and

$$xy = 2$$

are *nonlinear equations* if the variables are x and y.

An equation such as

$$\frac{x}{A} + y\,\mathrm{Log}(B) = p$$

is linear if the variables are x and y. On the other hand, if A and B are the variables, and x, y, and p are known values, then the equation is nonlinear.

In this chapter, we will consider several different methods for solving a set of linear equations. The solution of nonlinear equations is discussed in Chapters 8 and 10.

In cases where there are several unknowns and an equal number of independent equations, a unique solution is possible. The two linear equations

$$x - 2y = 1$$
$$2x + y = 7$$

represent two straight lines in the x-y plane. The simultaneous solution of these equations corresponds to the intersection of the two lines. Therefore, we can obtain a graphic solution by plotting the lines and finding the point of intersection.

The first equation has a y-intercept of $-0.5$ and a slope of 0.5. We can see this by rearranging the equation as

$$y = -0.5 + 0.5x$$

The second equation has an intercept of 7 and a slope of $-2$:

$$y = 7 - 2x$$

The two lines intersect at the location

$$x = 3, \qquad y = 1$$

which represents the solution to the problem. We can verify the solution by substituting these values for the variables in the original equations.

Another method for finding the simultaneous solution to these two equations is to multiply the second equation by 2 and add it to the first equation. The resulting equation contains only the variable x, and so it can be solved directly.

$$
\begin{array}{rcl}
x - 2y & = & 1 \\
\underline{4x + 2y} & = & \underline{14} \\
5x & = & 15 \quad \text{or} \quad x = 3
\end{array}
$$

We can then substitute this value of x into either one of the original equations to find the corresponding value of y.

Both of the above methods are suitable for solving two simultaneous equations. However, in general, these methods are tedious for larger numbers of equations. In the next section, we will study a more sophisticated method of solving simultaneous equations.

## *Solution by Cramer's Rule*

A technique known as *Cramer's rule* is useful for solving two or three simultaneous equations. This method, which uses matrices and vectors, is particularly suitable for solving linear equations by hand or with a pocket calculator. Using this approach, we write the equations with the unknowns on one side of the equal sign and the constant terms on the other. Terms containing the same unknowns are aligned vertically. The two equations in the previous section were initially written this way.

We place the coefficients of the unknowns in a form known as the *coefficient matrix*. The constant terms are put into a separate vector. For our two equations, this produces the matrix A and the vector Z.

$$
A = \begin{bmatrix} 1 & -2 \\ 2 & 1 \end{bmatrix} \qquad Z = \begin{bmatrix} 1 \\ 7 \end{bmatrix}
$$

The solution is found from the relationship

$$x = \frac{D_1}{D} \qquad y = \frac{D_2}{D}$$

where D is the determinant of the coefficient matrix

$$D = \begin{vmatrix} 1 & -2 \\ 2 & 1 \end{vmatrix} = (1)(1) - (2)(-2) = 5$$

The determinant $D_1$ is found by substituting the constant vector into column 1 of the matrix. This column corresponds to the unknown x:

$$D_1 = \begin{vmatrix} 1 & -2 \\ 7 & 1 \end{vmatrix} = (1)(1) - (7)(-2) = 15$$

Similarly, $D_2$ is obtained by substituting the constant vector into column 2:

$$D_2 = \begin{vmatrix} 1 & 1 \\ 2 & 7 \end{vmatrix} = (1)(7) - (2)(1) = 5$$

The solution is then

$$x = \frac{15}{5} = 3 \qquad \text{and} \qquad y = \frac{5}{5} = 1$$

Of course, some sets of equations have no unique solutions. We will now examine how Cramer's rule handles such equations.

## *Linear Dependence*

If the determinant of the coefficient matrix is zero, then no unique solution is possible. This will occur whenever there is a *linear dependence* among the equations, that is, one of the equations can be obtained from a combination of the others. Then the matrix is said to be *singular*. As an example of linear dependence, consider the two equations

$$x + \phantom{2}y = 5$$
$$2x + 2y = 2$$

The coefficient matrix for these two equations is

$$\begin{bmatrix} 1 & 1 \\ 2 & 2 \end{bmatrix}$$

and the corresponding determinant is zero. These equations represent parallel lines. Because these lines do not intersect, there can be no solution.

As a second example, consider the two equations

$$x + y = 5$$
$$2x + 2y = 10$$

This example produces the same singular coefficient matrix as the previous one. Now, however, the two lines lie on top of each other, and so there are an infinite number of solutions.

## *Example: A Direct-Current Electrical Circuit*

Now that we have a powerful method of solving two or three simultaneous equations, let us consider a practical application. An interesting example that requires the solution of simultaneous equations is provided by an electrical circuit. Consider, for example, the network of resistors and direct-current (DC) voltage sources shown in Figure 4.1. This

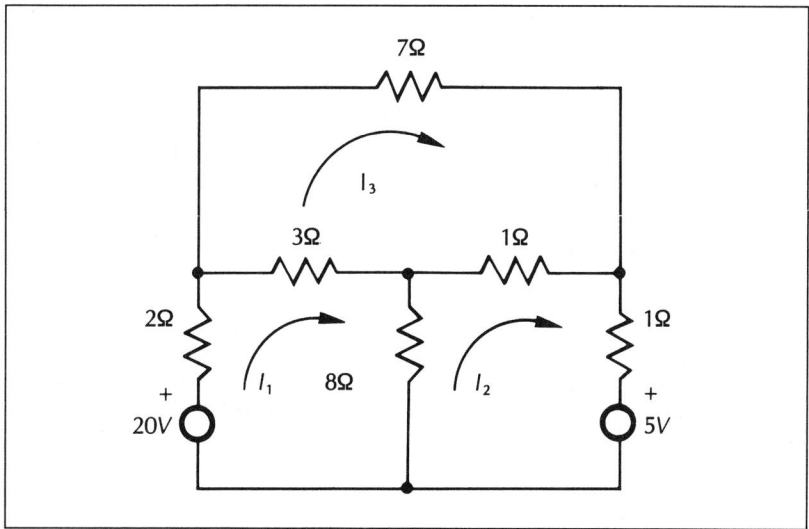

**Figure 4.1:** A network of resistors and DC sources.

circuit has four nodes and six branches. There is a 20-volt source on the left side of the network and a 5-volt source on the right side. In addition, there are six resistors of known value.

The problem is to find the resulting branch currents and the voltages across each resistor. We can determine the electrical currents in the six separate branches by solving six simultaneous equations. However, we can simplify the problem by considering three loop currents and the corresponding three simultaneous equations. Then we can calculate the branch currents from the loop currents.

The lower-left loop current is designated as $I_1$, the lower-right loop current is $I_2$, and the upper loop current is $I_3$. Then the three loop equations are derived from the Kirchhoff voltage law by going around each loop in a counter-clockwise direction:

$$13I_1 - 8I_2 - 3I_3 - 20 = 0 \text{ (lower-left loop)}$$

$$-8I_1 + 10I_2 - I_3 + 5 = 0 \text{ (lower-right loop)}$$

$$-3I_1 - I_2 + 11I_3 = 0 \text{ (upper loop)}$$

The corresponding coefficient matrix and constant vector are

$$\begin{bmatrix} 13 & -8 & -3 \\ -8 & 10 & -1 \\ -3 & -1 & 11 \end{bmatrix} \begin{bmatrix} 20 \\ -5 \\ 0 \end{bmatrix}$$

Proceeding by Cramer's rule, we find

$$D = \begin{vmatrix} 13 & -8 & -3 \\ -8 & 10 & -1 \\ -3 & -1 & 11 \end{vmatrix} \quad D_1 = \begin{vmatrix} 20 & -8 & -3 \\ -5 & 10 & -1 \\ 0 & -1 & 11 \end{vmatrix}$$

$$D_2 = \begin{vmatrix} 13 & 20 & -3 \\ -8 & -5 & -1 \\ -3 & 0 & 11 \end{vmatrix} \quad D_3 = \begin{vmatrix} 13 & -8 & 20 \\ -8 & 10 & -5 \\ -3 & -1 & 0 \end{vmatrix}$$

We can find the determinants for these four matrices by using the DETERM program, given in Listing 3.3 of the previous chapter. The resulting loop currents are

$$I_1 = \frac{D_1}{D} = \frac{1725}{575} = 3 \text{ amps}$$

$$I_2 = \frac{D_2}{D} = \frac{1150}{575} = 2 \text{ amps}$$

$$I_3 = \frac{D_3}{D} = \frac{575}{575} = 1 \text{ amp}$$

Although the DETERM program gives us the correct answer, it is unnecessarily complicated. We must enter 9 numbers into the computer program for each of the 4 determinants, for a total of 36 separate values.

We will now study an efficient program designed specifically for solving simultaneous equations by Cramer's rule.

## A More Elegant Use of Cramer's Rule

The program shown in Listing 4.1 considerably simplifies the process. The mathematical operations are the same as in our first solution, but we only need to enter 12 numbers: 9 for the coefficients and 3 for the constant vector.

```
PROGRAM Simq1;
{ Pascal program to solve three }
{ simultaneous equations by Cramer's rule }

CONST
   Rmax = 3;
   Cmax = 3;

TYPE
   Arys    = ARRAY[1..Cmax] OF Real;
   Ary2s   = ARRAY[1..Rmax, 1..Cmax] OF Real;

VAR
   Y, Coef: Arys;
        A: Ary2s;
        N: Integer;
    YesNo: Char;
    Error: Boolean;

PROCEDURE Get_Data(VAR A: Ary2s;
                   VAR Y: Arys;
                   VAR N: Integer);

{ Get values for N and arrays A, Y }
```

**Listing 4.1:** Solution of three linear equations by Cramer's rule

```
VAR
  I, J: Integer;

BEGIN  { procedure Get_Data }
  WriteLn;
  N := Rmax;
  FOR I := 1 TO N DO
    BEGIN
      WriteLn(' Equation ', I:3);
      FOR J := 1 TO N DO
        BEGIN
          Write(J:3, ': ');
          Read(A[I,J])
        END;
      Write(', C: ');
      ReadLn(Y[I])
    END;
  WriteLn;
  FOR I:= 1 TO N  DO
    BEGIN
      FOR J:= 1 TO N DO
        Write(A[I,J]:7:4, '  ');
      WriteLn(' : ', Y[I]:7:4)
    END;
  WriteLn
END; { procedure Get_Data }

PROCEDURE Write_Data;
{ print out the answers }

VAR
  I: Integer;

BEGIN  { Write_Data }
  FOR I := 1 TO N DO
    Write(Coef[I]:9:5);
  WriteLn
END; { Write_Data }

PROCEDURE Solve(A: Ary2s;
                Y: Arys;
           VAR  Coef: Arys;
                N: Integer;
           VAR  Error: Boolean);

VAR
    B: Ary2s;
  Det: Real;
  I, J: Integer;

FUNCTION Deter(A: Ary2s): Real;
{ the determinant of a 3-by-3 matrix }

VAR
  Sum: Real;
```

**Listing 4.1:** Solution of three linear equations by Cramer's rule (continued).

```
BEGIN    { function Deter }
  Sum := A[1,1]*(A[2,2]*A[3,3]- A[3,2]*A[2,3])
       - A[1,2]*(A[2,1]*A[3,3]- A[3,1]*A[2,3])
       + A[1,3]*(A[2,1]*A[3,2]- A[3,1]*A[2,2]);
  Deter := Sum
END; { function Deter }

PROCEDURE Setup(VAR B: Ary2s;
           VAR  Coef: Arys;
                    J: Integer);
VAR
  I: Integer;

BEGIN    { Setup }
  FOR I := 1 TO N DO
    BEGIN
      B[I,J] := Y[I];
      IF J > 1 THEN B[I,J-1] := A[I,J-1]
    END;
  Coef[J] := Deter(B) / Det
END; { Setup }

BEGIN  { procedure solve }
  Error := False;
  FOR I := 1 TO N DO
    FOR J := 1 TO N DO
      B[I,J] := A[I,J];
  Det := Deter(B);
  IF Det = 0.0 THEN
    BEGIN
      Error := True;
      WriteLn(' ERROR: matrix singular')
    END
  ELSE
    BEGIN
      Setup(B, Coef, 1);
      Setup(B, Coef, 2);
      Setup(B, Coef, 3)
    END  { ELSE. }
END; { procedure solve }

BEGIN   { main program }
  WriteLn;
  WriteLn
    (' Simultaneous solution by Cramer''s rule');
  REPEAT
    Get_Data(A, Y, N);
    Solve(A, Y, Coef, N, Error);
    IF NOT Error THEN  Write_Data;
    WriteLn;
    Write(' More? ');
    ReadLn(YesNo)
  UNTIL (UpCase(YesNo) <> 'Y')
END.
```

**Listing 4.1:** Solution of three linear equations by Cramer's rule (continued).

### Running the Cramer's Rule Program

Type the program shown in Listing 4.1, give it the name SIMQ1, and run it. In this program, you enter the coefficients and constant term for the first equation on the first line, the corresponding values for the second equation on the second line, and the data for the third equation on the third line. Press the Enter key after you enter each coefficient or constant term. The video screen will look like this:

```
Simultaneous solution by Cramer's rule
Equation   1
    1:   13    2:  −8    3:  −3 ,   C:   20
Equation   2
    1:  −8    2:   10    3:  −1 ,   C:  −5
Equation   3
    1:  −3    2:  −1    3:  11 ,   C:   0
```

The data will be displayed again, and the solution will follow:

```
    13.0000   −8.0000   −3.0000   : 20.0000
   −8.0000    10.0000   −1.0000   : −5.0000
   −3.0000   −1.0000    11.0000   : 0.0000

    3.00000    2.00000    1.00000
```

Then you will see the question

```
More?
```

Type Y and press Enter if you want to solve another set. Otherwise, type N or press Ctrl-C.

# Solution by Gauss Elimination

Cramer's rule is an effective method for solving two or three simultaneous equations, but the computation time increases with the fourth power of the matrix size. Consequently, it takes about 16 times longer to solve six simultaneous equations than it does to solve three of them. The *Gauss elimination* method is more efficient for solving four or more simultaneous equations. The computation time increases with the third power of the number of equations, and so it takes about eight times longer to solve six equations than it does to solve three.

The Gauss elimination method is most suitable for computer program calculations because it is fairly complicated. It involves frequent multiplication, division, and subtraction operations, and these contribute to a loss of precision. Thus, this method is not suitable for simultaneously solving many equations, say more than 15. (At the end of this chapter, we will discuss a method for solving a large number of equations.) Even though you won't be using this method for hand calculations, we'll go through it step by step to explain how it works.

## *The Steps of the Gauss Elimination Method*

Using the Gauss elimination method, we manipulate the original equations so that the coefficient matrix contains a value of unity at each point on the major diagonal and zeros at each position below and to the left of the major diagonal.

Two basic types of matrix operations are used in the Gauss elimination method: scalar multiplication and addition. Any equation can be multiplied by a constant without changing the result. This is equivalent to multiplying one row of the coefficient matrix and the corresponding element in the constant vector by the same value. Also, any equation can be replaced by the sum of two equations.

The following steps will demonstrate the use of Gauss elimination to solve the equations that we derived from the electric circuit shown in Figure 4.1. Initially, the coefficient matrix and the constant vector are

$$\begin{bmatrix} 13 & -8 & -3 \\ -8 & 10 & -1 \\ -3 & -1 & 11 \end{bmatrix} \quad \begin{bmatrix} 20 \\ -5 \\ 0 \end{bmatrix}$$

The first variable is eliminated from all but the first equation. We divide each element in the first row by the first element in the row (the pivot element). This generates the value of unity in the first diagonal position. By this means, the first equation becomes

$$[1 \quad -0.61 \quad -0.23] \quad [1.5]$$

The first unknown is eliminated from the second row by combining the first two rows. We multiply the new first row by the first element in the second row, then subtract the first row from the second row. The new second row is

$$[0 \quad 5.1 \quad -2.8] \quad [7.3]$$

In a similar fashion, the first variable is eliminated from the third equation. The three equations are now approximately

$$\begin{bmatrix} 1 & -0.61 & -0.23 \\ 0 & 5.1 & -2.8 \\ 0 & -2.8 & 10.3 \end{bmatrix} \begin{bmatrix} 1.5 \\ 7.3 \\ 4.6 \end{bmatrix}$$

The next step produces the value of unity in the second position of the second row (the new pivot). We divide the second line by the second element. The second variable is eliminated from the third equation by generating a zero in the second position of this row, just under the pivot element. The three equations now look like this

$$\begin{bmatrix} 1 & -0.61 & -0.23 \\ 0 & 1 & -0.56 \\ 0 & 0 & 8.7 \end{bmatrix} \begin{bmatrix} 1.5 \\ 1.4 \\ 8.7 \end{bmatrix}$$

The final step of this phase is to obtain a value of unity in the third pivot position. This is done by dividing the third equation by the pivot value. The matrix and constant vector are now

$$\begin{bmatrix} 1 & -0.61 & -0.23 \\ 0 & 1 & -0.56 \\ 0 & 0 & 1 \end{bmatrix} \begin{bmatrix} 1.5 \\ 1.4 \\ 1 \end{bmatrix}$$

This corresponds to the three equations

$$x - 0.61y - 0.23z = 1.5$$
$$y - 0.56z = 1.4$$
$$z = 1$$

The third equation can be solved directly since it has only one unknown. The result is

$$z = 1$$

The second equation can be written as:

$$y = 1.4 + 0.56z$$

By substituting the value of z into this equation, we find the value of y to be 2. Substituting the values of y and z into the first equation produces the value of 3 for x. This phase of the calculations is called *back substitution*.

## Improving the Accuracy
## of the Gauss Elimination Method

We can improve the accuracy of the Gauss elimination method by interchanging two rows so that the element with the largest absolute magnitude becomes the pivot element. For example, suppose that the previous three equations had been originally written in this order:

$$\begin{bmatrix} -3 & -1 & 11 \\ 13 & -8 & -3 \\ -8 & 10 & -1 \end{bmatrix} \quad \begin{bmatrix} 0 \\ 20 \\ -5 \end{bmatrix}$$

Then the top equation could be divided by $-3$ to make the pivot value unity. However, the result will be more accurate if we first interchange the first and second rows. This will put the larger value of 13 in the pivot position. After the first variable is eliminated from the second and third equations, the result is

$$\begin{bmatrix} 1 & -0.61 & -0.23 \\ 0 & -2.8 & 10.3 \\ 0 & 5.1 & -2.8 \end{bmatrix} \quad \begin{bmatrix} 1.5 \\ 4.6 \\ 7.3 \end{bmatrix}$$

Again, we can increase the precision by interchanging equations 2 and 3 to put the larger element of 5.1 in the second pivot position.

You may have to interchange rows for another reason. If a value of zero appears on the major diagonal, then it is not possible to divide the row by this pivot element. Thus, you would have to interchange this row with the one below it to remove the zero from the pivot position.

Now that we have gone through the rather tedious process of using the Gauss elimination method to solve equations by hand, we can appreciate the elegance of the Turbo Pascal program presented in the following section.

## Programming the Gauss Elimination Method

The program shown in Listing 4.2 can simultaneously solve a set of linear equations by using the Gauss elimination method. The program is designed for a maximum of eight equations. To solve a larger number of equations, change the size of the two variables Maxr and Maxc near the beginning of the program.

The programs in this chapter are closely related. Therefore, you can create each new version by altering a copy of a previous program. As a consequence, there may be features in a program that are not used until

```
PROGRAM Gaus;
{ Turbo Pascal program to perform }
{ simultaneous solution by Gaussian elimination }
{ procedure Gauss is included }

CONST
  Maxr = 8;
  Maxc = 8;

TYPE
  Ary   = ARRAY[1..Maxr] OF Real;
  Arys  = ARRAY[1..Maxc] OF Real;
  Ary2s = ARRAY[1..Maxr, 1..Maxc] OF Real;

VAR
  Y, Coef: Arys;
        A: Ary2s;
     N, M: Integer;
    Error: Boolean;

PROCEDURE Get_Data(VAR A: Ary2s;
                   VAR Y: Arys;
                   VAR N, M: Integer);
{ Get values for N and arrays A, Y }

VAR
  I, J: Integer;

BEGIN
  WriteLn;
  REPEAT
    Write(' How many equations? ');
    ReadLn(N);
    M := N
  UNTIL N < Maxr;
  IF N > 1 THEN
    BEGIN
      FOR I := 1 TO N DO
        BEGIN
          WriteLn(' Equation', I:3);
          FOR J := 1 TO N DO
            BEGIN
              Write(J:3, ': ');
              Read(A[I,J])
            END;
          Write(', C: ');
          ReadLn(Y[I]);
        END;
      WriteLn;
      FOR I:= 1 TO N DO
        BEGIN
          FOR J:= 1 TO M DO
            Write(A[I,J]:7:4, ' ');
          WriteLn(' : ', Y[I]:7:4)
        END;
```

**Listing 4.2:** Simultaneous solution by Gauss elimination.

```
          WriteLn
      END  { if N>1 }
END; { procedure Get_Data }

PROCEDURE Write_Data;
{ print out the answers }

VAR
  I: Integer;

BEGIN
  FOR I := 1 TO M DO
    Write(Coef[I]:9:5);
  WriteLn
END; { Write_Data }

PROCEDURE Gauss
              (A: Ary2s;
               Y: Arys;
          VAR Coef: Arys;
              Ncol: Integer;
          VAR Error: Boolean);
{ matrix solution by Gaussian Elimination }
{ Adapted from Gilder }

VAR
  B: Ary2s; { work array, Nrow,Ncol }
  W: Arys;  { work array, Ncol long }
  I, J, I1, K, L, N: Integer;
  Hold, Sum, T, Ab, Big: Real;

BEGIN
  Error := False;
  N := Ncol;
  FOR I := 1 TO N DO
    BEGIN { copy to work arrays }
      FOR J := 1 TO N DO
        B[I,J] := A[I,J];
      W[I] := Y[I]
    END;
  FOR I := 1 TO N - 1 DO
    BEGIN
      Big := Abs(B[I,I]);
      L := I;
      I1 := I + 1;
      FOR J := I1 TO N DO
        BEGIN { search for largest element }
          Ab := Abs(B[J,I]);
          IF Ab > Big THEN
            BEGIN
              Big := Ab;
              L := J
            END
        END;
      IF Big = 0.0 THEN
        Error := True
      ELSE
```

**Listing 4.2:** Simultaneous solution by Gauss elimination (continued).

```
           BEGIN
             IF L <> I THEN
               BEGIN
                 { interchange rows to put }
                 { largest element on diagonal }
                 FOR J := 1 TO N DO
                   BEGIN
                     Hold := B[L,J];
                     B[L,J] := B[I,J];
                     B[I,J] := Hold
                   END;
                 Hold := W[L];
                 W[L] := W[I];
                 W[I] := Hold
               END; { if L <> I }
             FOR  J := I1 TO N DO
               BEGIN
                 T := B[J,I] / B[I,I];
                 FOR K := I1 TO N DO
                   B[J,K] := B[J,K] - T * B[I,K];
                 W[J] := W[J] - T * W[I]
               END   { J loop }
           END    { if Big }
      END; { I loop }
  IF B[N,N] = 0.0 THEN
    Error := True
  ELSE
    BEGIN
      Coef[N] := W[N] / B[N,N];
      I := N - 1;
      { back substitution }
      REPEAT
        Sum := 0.0;
        FOR J := I+1 TO N DO
          Sum := Sum + B[I,J] * Coef[J];
        Coef[I] := (W[I] - Sum) / B[I,I];
        I := I - 1
      UNTIL I = 0
    END; { IF B[N,N] = 0 }
  IF Error THEN
    WriteLn('ERROR: Matrix singular ')
END; { Gauss }

BEGIN { main program }
  WriteLn;
  WriteLn
    (' Simultaneous solution by Gauss elimination');
  REPEAT
    Get_Data(A, Y, N, M);
    IF N > 1 THEN
      BEGIN
        Gauss(A, Y, Coef, N, Error);
        IF NOT Error THEN  Write_Data
      END
  UNTIL N < 2
END.
```

**Listing 4.2:** Simultaneous solution by Gauss elimination (continued).

later. For example, you can see that the symbol m is not needed in procedure Get_Data of Listing 4.2. However, this symbol is used in the program shown in Listing 4.7. The idea here is to designate the number of rows with the symbol n and the number of columns with the symbol m. Thus, m equals n, except in Listing 4.7.

## *Running the Gauss Elimination Program*

Type the program shown in Listing 4.2, give it the name GAUSS.PAS, and then run it. This program is similar to the previous one, except that you first enter the number of equations to be solved. Then you enter the coefficients for each equation and the corresponding constant vectors, one equation on each line. Press the Enter key after you type each number. Solve the set of equations we considered earlier in this chapter, and verify that the solution vector is [3  2  1]. The results should look like Figure 4.2.

After displaying the solution, the program begins again. This time find the solution to the following three equations:

$$\begin{bmatrix} 1 & 1 & 1 \\ 2 & 1 & -1 \\ 3 & 1 & -3 \end{bmatrix} \begin{bmatrix} 6 \\ 1 \\ -4 \end{bmatrix}$$

```
                 Simultaneous solution by Gauss elimination

                 How many equations? 3
                 Equation  1
                  1: 13    2: -8    3: -3 , C: 20
                 Equation  2
                  1: -8    2: 10    3: -1 , C: -5
                 Equation  3
                  1: -3    2: -1    3: 11 , C: 0

                 13.0000   -8.0000   -3.0000   : 20.0000
                 -8.0000   10.0000   -1.0000   : -5.0000
                 -3.0000   -1.0000   11.0000   :  0.0000

                  3.00000  2.00000  1.00000
```

**Figure 4.2:** Output: solution of the electrical circuit problem using Gauss elimination.

If you try the values $x = 1$, $y = 2$, and $z = 3$ in these equations, you will see that they satisfy all the three equations. However, the program will report

ERROR: matrix singular

The problem is that one of the equations is a linear combination of the other two. In this example, the third equation is equal to twice the second equation minus the first equation. This makes the coefficient matrix singular.

While Turbo Pascal readily discovers the linear dependence, many Pascal, Fortran, and BASIC compilers cannot find the problem. Although the determinant of the coefficient matrix is small, it may not equal zero because of roundoff errors that accumulate during the elimination process. Then different algorithms for solving simultaneous equations may give different answers. Therefore, if there is any question of linear dependence, you should use at least two different techniques to solve the equations.

For the third task, enter two lines (equations) that are the same, for example

$$\begin{bmatrix} 1 & 1 & 1 \\ 1 & 1 & 1 \\ 2 & 1 & -1 \end{bmatrix} \begin{bmatrix} 6 \\ 6 \\ 1 \end{bmatrix}$$

Again, the Gauss elimination program will report a singular matrix. Terminate the program by entering a zero or pressing Ctrl-C.

## Solution by Gauss-Jordan Elimination

A variation of the Gauss elimination method is the *Gauss-Jordan elimination method*. This approach shares most of the advantages and disadvantages of the Gauss elimination technique.

Execution time for this method is a third-order function of the matrix size, and there are many multiplication, division, and subtraction operations that contribute to loss of accuracy. Furthermore, the Gauss-Jordan algorithm is more complicated than the one for Gauss elimination. Nevertheless, the Gauss-Jordan elimination technique will generally be the most useful to us because the inverse of the coefficient matrix is readily obtained along with the solution vector. Therefore, we will use it in later chapters of this book. Let us examine how this method works.

## *Details of the Gauss-Jordan Elimination Method*

In the Gauss-Jordan elimination method, the elements of the major diagonal are converted to unity as with the Gauss elimination method. However, next the elements both above and below the major diagonal are converted to zeros. Thus, the coefficient matrix becomes a unit matrix, and the constant vector becomes the solution vector. For the Gauss-Jordan solution of the electrical circuit given in Figure 4.1, the final set of numbers is

$$
\begin{bmatrix} 1 & 0 & 0 \\ 0 & 1 & 0 \\ 0 & 0 & 1 \end{bmatrix}
\begin{bmatrix} 3 \\ 2 \\ 1 \end{bmatrix}
$$

This corresponds to the three equations:

$$
\begin{aligned}
x &= 3 \\
y &= 2 \\
z &= 1
\end{aligned}
$$

for which the solution is

$$ x = 3, \quad y = 2, \quad z = 1 $$

Suppose that a unit matrix is initially placed to the right of the original set of equations. If all the operations are done on this unit matrix as they are on the other matrix and vector, then it will be converted into the inverse of the original coefficient matrix at the conclusion of the calculation. That is, the set

$$
\begin{bmatrix} 13 & -8 & -3 \\ -8 & 10 & -1 \\ -3 & -1 & 11 \end{bmatrix}
\begin{bmatrix} 20 \\ -5 \\ 0 \end{bmatrix}
\begin{bmatrix} 1 & 0 & 0 \\ 0 & 1 & 0 \\ 0 & 0 & 1 \end{bmatrix}
$$

will be transformed into

$$
\begin{bmatrix} 1 & 0 & 0 \\ 0 & 1 & 0 \\ 0 & 0 & 1 \end{bmatrix}
\begin{bmatrix} 3 \\ 2 \\ 1 \end{bmatrix}
\begin{bmatrix} 0.190 & 0.158 & 0.066 \\ 0.158 & 0.233 & 0.064 \\ 0.066 & 0.064 & 0.115 \end{bmatrix}
$$

Then the matrix on the right is the inverse of the original coefficient matrix.

In the previous example, we show the inverted matrix separately from the original matrix. However, it is not necessary to use two separate matrices for these operations. We can generate the inverse matrix in the same space occupied by the original coefficient matrix. At the conclusion of the operation, the inverse matrix will appear in place of the coefficient matrix. Of course, the solution vector will appear in place of the original constant vector. Thus, if the coefficient matrix and constant vector

$$
\begin{bmatrix} 13 & -8 & -3 \\ -8 & 10 & -1 \\ -3 & -1 & 11 \end{bmatrix} \quad \begin{bmatrix} 20 \\ -5 \\ 0 \end{bmatrix}
$$

are given to the Gauss-Jordan elimination routine, the matrix inverse and solution vector can be put into the same array space:

$$
\begin{bmatrix} 0.190 & 0.158 & 0.066 \\ 0.158 & 0.233 & 0.064 \\ 0.066 & 0.064 & 0.115 \end{bmatrix} \quad \begin{bmatrix} 3 \\ 2 \\ 1 \end{bmatrix}
$$

We will now present a Turbo Pascal program that accomplishes this rather complex algorithm.

## Programming the Gauss-Jordan Elimination Method

The program shown in Listing 4.3 will solve simultaneous linear equations by using the Gauss-Jordan elimination method. The main program calls procedure Gaussj, which is given as a separate procedure in Listing 4.4. We will use procedure Gaussj in several programs in this book.

This new version copies the original coefficient matrix A into matrix B before calling procedure Gaussj. Gaussj then returns the inverse of the coefficient matrix in place of the original matrix B. In a similar way, the data vector, Y, is isolated from the solution vector, Coef, calculated by Gaussj.

```
PROGRAM Solvgj;
{ Turbo Pascal program to perform simultaneous }
{ solution by Gauss-Jordan elimination }

CONST
  Maxr = 8;
  Maxc = 8;
```

**Listing 4.3:** Solution of simultaneous equations by Gauss-Jordan elimination.

```
TYPE
  Ary   = ARRAY[1..Maxr] OF Real;
  Arys  = ARRAY[1..Maxc] OF Real;
  Ary2s = ARRAY[1..Maxr, 1..Maxc] OF Real;

VAR
    Y, Coef: Arys;
       A, B: Ary2s;
 N, M, I, J: Integer;
      Error: Boolean;


PROCEDURE Get_Data(VAR A: Ary2s;
                   VAR Y: Arys;
                   VAR N, M: Integer);
{ Get values for N and arrays A, Y }

VAR
  I, J: Integer;

BEGIN
  WriteLn;
  REPEAT
    Write(' How many equations? ');
    ReadLn(N);
    M := N
  UNTIL N < Maxr;
  IF N > 1 THEN
    BEGIN
      FOR I := 1 TO N DO
        BEGIN
          WriteLn(' Equation', I:3);
          FOR J := 1 TO N DO
            BEGIN
              Write(J:3, ': ');
              Read(A[I,J])
            END;
          Write(', C: ');
          ReadLn(Y[I])   { clear line }
        END;
      WriteLn;
      FOR I:= 1 TO N  DO
        BEGIN
          FOR J:= 1 TO M DO
            Write(A[I,J]:7:4, '  ');
          WriteLn(' : ', Y[I]:7:4)
        END;
      WriteLn
    END { if N>1 }
END; { procedure Get_Data }

PROCEDURE Write_Data;
{ print out the answers }

VAR
  I: Integer;
```

**Listing 4.3:** Solution of simultaneous equations by Gauss-Jordan elimination (continued).

```
BEGIN
  FOR I := 1 TO M DO
    Write(Coef[I]:9:5);
  WriteLn
END; { Write_Data }

{$I GAUSSJ.PAS} {Listing 4.4}

BEGIN  { main program }
  WriteLn;
  WriteLn
    (' Simultaneous solution by Gauss-Jordan elimination');
  REPEAT
    Get_Data(A, Y, N, M);
    IF N > 1 THEN
      BEGIN
        FOR I := 1 TO N DO
          FOR J := 1 TO N DO
            B[I,J] := A[I,J]; { Setup work array }
        Gaussj(B, Y, Coef, N, Error);
        IF NOT Error THEN Write_Data
      END
  UNTIL N < 2
END.
```

**Listing 4.3:** Solution of simultaneous equations by Gauss-Jordan elimination (continued).

```
PROCEDURE Gaussj
      (VAR B: Ary2s;   { square matrix of coefficients }
           Y: Arys;    { constant vector }
      VAR Coef: Arys;  { solution vector }
         Ncol: Integer;   { order of matrix }
      VAR Error: Boolean); { True if matrix singular }

{ Gauss Jordan matrix inversion and solution
  Adapted from McCormick
    B(N,N) coefficient matrix, becomes inverse
    Y(N)   original constant vector
    W(N,M) constant vector(s) become solution vector
    Determ is the determinant
    Error = 1 if singular
    Index(N,3)
    Nv is number of constant vectors }

VAR
      W: ARRAY[1..Maxc, 1..Maxc] OF Real;
  Index: ARRAY[1..Maxc, 1..3] OF Integer;
  I, J, K, L, Nv, Irow, Icol, N, L1   : Integer;
  Determ, Pivot, Hold, Sum, T, Ab, Big: Real;
```

**Listing 4.4:** The Gauss-Jordan elimination procedure.

```
PROCEDURE Swap(VAR A, B: Real);

VAR
  Hold: Real;

BEGIN  { Swap }
  Hold := A;
  A := B;
  B := Hold
END; { procedure Swap }

BEGIN     { Gauss-Jordan main program }
  Error := False;
  Nv := 1; { single constant vector }
  N := Ncol;
  FOR I := 1 TO N DO
    BEGIN
      W[I, 1] := Y[I]; { copy constant vector }
      Index[I, 3] := 0
    END;
  Determ := 1.0;

  FOR I := 1 TO N DO
    BEGIN
      { search for largest element }
      Big := 0.0;
      FOR J := 1 TO N DO
        BEGIN
          IF Index[J, 3] <> 1 THEN
            BEGIN
              FOR K := 1 TO N DO
                BEGIN
                  IF Index[K, 3] > 1 THEN
                    BEGIN
                      WriteLn(' ERROR: matrix singular');
                      Error := True;
                      Exit  { procedure Gaussj }
                    END;
                  IF Index[K, 3] < 1 THEN
                    IF Abs(B[J, K]) > Big THEN
                      BEGIN
                        Irow := J;
                        Icol := K;
                        Big := Abs(B[J, K])
                      END
                END { K loop }
            END
        END; { J loop }
      Index[Icol, 3] := Index[Icol, 3] + 1;
      Index[I, 1] := Irow;
      Index[I, 2] := Icol;

      { interchange rows to put pivot on diagonal }
      IF Irow <> Icol THEN
        BEGIN
          Determ := - Determ;
```

**Listing 4.4:** The Gauss-Jordan elimination procedure (continued).

```
     FOR L := 1 TO N DO
        Swap(B[Irow, L], B[Icol, L]);
     IF Nv > 0 THEN
        FOR L := 1 TO Nv DO
           Swap(W[Irow, L], W[Icol, L])
  END; { if Irow <> Icol }

     { divide Pivot row by Pivot column }
     Pivot := B[Icol, Icol];
     Determ := Determ * Pivot;
     B[Icol, Icol] := 1.0;
     FOR L := 1 TO N DO
        B[Icol, L] := B[Icol, L] / Pivot;
     IF Nv > 0 THEN
        FOR L := 1 TO Nv DO
           W[Icol, L] := W[Icol, L] / Pivot;

        { reduce nonpivot rows }
        FOR L1 := 1 TO N DO
           BEGIN
              IF L1 <> Icol THEN
                 BEGIN
                    T := B[L1, Icol];
                    B[L1, Icol] := 0.0;
                    FOR L := 1 TO N DO
                       B[L1, L] := B[L1, L] - B[Icol, L] * T;
                    IF Nv > 0 THEN
                       FOR L := 1 TO Nv DO
                          W[L1, L] := W[L1, L] - W[Icol, L] * T;
                 END    { IF L1 <> Icol }
           END
     END; { I loop }
     { interchange columns }
     FOR I := 1 TO N DO
        BEGIN
           L := N - I + 1;
           IF Index[L, 1] <> Index[L, 2] THEN
              BEGIN
                 Irow := Index[L, 1];
                 Icol := Index[L, 2];
                 FOR K := 1 TO N DO
                    Swap(B[K, Irow], B[K, Icol])
              END { if Index }
        END; { I loop }
     FOR K := 1 TO N DO
        IF Index[K, 3] <> 1 THEN
           BEGIN
              WriteLn(' ERROR: matrix singular');
              Error := True;
              Exit  { procedure Gaussj }
           END;
     FOR I := 1 TO N DO
        Coef[I] := W[I, 1];
  END; { procedure Gaussj }
```

**Listing 4.4:** The Gauss-Jordan elimination procedure (continued).

### *Running the Gauss-Jordan Elimination Program*

Type the main program shown in Listing 4.3, and give it the name SOLVGJ.PAS. You do not need to type the Gauss-Jordan elimination routine shown in Listing 4.4. On your original Turbo Pascal disk, you will find a program named either DEMO1-77.PAS or HILB.PAS. Copy it to your working disk with the name HILB.PAS. My Gauss-Jordan procedure is included in this file. Make a second copy in a separate disk file named GAUSSJ.PAS. Edit this copy to remove the extraneous parts, leaving only procedure Gaussj. Run SOLVGJ, and solve the electrical circuit equations

$$\begin{bmatrix} 13 & -8 & -3 \\ -8 & 10 & -1 \\ -3 & -1 & 11 \end{bmatrix} \quad \begin{bmatrix} 20 \\ -5 \\ 0 \end{bmatrix}$$

to verify that the solution is $[3 \quad 2 \quad 1]$.

We have seen how the Gauss-Jordan elimination program works for one coefficient matrix and its corresponding constant vector. Now we will explore ways of using and refining this program to solve systems that have multiple constant vectors. To illustrate this type of problem, we will continue with our circuit example. We will also continue our discussion of inverse coefficient matrices and how to use them in solving simultaneous equations.

## *Multiple Constant Vectors and Matrix Inversion*

In the previous chapter, we learned that we can solve a set of linear equations by multiplying the inverse of the coefficient matrix by the constant vector. For example, if the coefficient matrix is A and the constant vector is Y, then the solution vector B is $B = A^{-1}Y$. However, the methods developed in this chapter (Cramer's rule, Gauss elimination, and Gauss-Jordan elimination) obtain the solution to a set of linear equations by direct methods rather than by using the inverse of the coefficient matrix.

Sometimes we need to solve several sets of simultaneous equations that have the same coefficient matrix but different constant vectors. As we learned earlier, we could invert the coefficient matrix and obtain the separate solution vectors from the product of the inverted matrix and each constant vector. Even with this example, however, it is generally faster to use the Gauss-Jordan elimination technique on the matrix and all the constant vectors simultaneously. In fact, the Gauss-Jordan procedure given in the previous section is already programmed for multiple constant vectors.

Consider again the electrical circuit shown in Figure 4.1. Suppose that we would like to determine the loop currents for three different circuit configurations:

- The original circuit

- The circuit with the 5-volt source reversed

- The circuit with both voltage sources reversed

These three different arrangements correspond to the following three sets of equations:

$$\begin{bmatrix} 13 & -8 & -3 \\ -8 & 10 & -1 \\ -3 & -1 & 11 \end{bmatrix} \begin{bmatrix} 20 \\ -5 \\ 0 \end{bmatrix}$$

$$\begin{bmatrix} 13 & -8 & -3 \\ -8 & 10 & -1 \\ -3 & -1 & 11 \end{bmatrix} \begin{bmatrix} 20 \\ 5 \\ 0 \end{bmatrix}$$

$$\begin{bmatrix} 13 & -8 & -3 \\ -8 & 10 & -1 \\ -3 & -1 & 11 \end{bmatrix} \begin{bmatrix} -20 \\ 5 \\ 0 \end{bmatrix}$$

We could solve all three sets of equations separately using one of the preceding techniques. However, since these equations all have the same coefficient matrix, it is more efficient to solve all three conditions simultaneously by using a variation of our Gauss-Jordan procedure.

The coefficient matrix contains the values common to the three different circuits. However, the constant vector now becomes a *constant matrix*. Each column of the constant matrix represents a different circuit configuration and will produce the corresponding solution. Now the matrices entered into the Gauss-Jordan elimination routine are

$$\begin{bmatrix} 13 & -8 & -3 \\ -8 & 10 & -1 \\ -3 & -1 & 11 \end{bmatrix} \begin{bmatrix} 20 & 20 & -20 \\ -5 & 5 & 5 \\ 0 & 0 & 0 \end{bmatrix}$$

where the first matrix is the usual coefficient matrix and the second has the constant vectors.

The Gauss-Jordan elimination routine returns the following values to the calling program:

$$\begin{bmatrix} 0.190 & 0.158 & 0.066 \\ 0.158 & 0.233 & 0.064 \\ 0.066 & 0.064 & 0.115 \end{bmatrix} \quad \begin{bmatrix} 3 & 4.58 & -3 \\ 2 & 4.33 & -2 \\ 1 & 1.64 & -1 \end{bmatrix}$$

The left matrix, which originally contained the coefficients, now has the inverse of the coefficient matrix. The right matrix, which initially contained the constant vectors, now has the corresponding solution vectors. Thus, for the three separate circuits, the answers are

|        | Circuit 1 | Circuit 2 | Circuit 3 |
|--------|-----------|-----------|-----------|
| $I_1$  | 3         | 4.58      | $-3$      |
| $I_2$  | 2         | 4.33      | $-2$      |
| $I_3$  | 1         | 1.64      | $-1$      |

In the previous version of the Gauss-Jordan elimination program, we copied the original coefficient matrix A into a work array B before calling procedure Gaussj. Also, at the beginning of the Gauss-Jordan elimination routine, the original constant vector Y was copied into the first column of the constant matrix W. Then, at the end of procedure Gaussj, the solution from the first column of matrix W was copied into the solution vector Coef. Now we will look at a refined version of this program.

# An Alternative Method for Programming Gauss-Jordan Elimination

An alternative version of the Gauss-Jordan elimination program is given in Listing 4.5. The parameters of the Gauss-Jordan procedure have been changed. As before, the procedure returns the inverse of the coefficient matrix in the array that held the original coefficient matrix. However, the solution matrix is now returned in the array that originally held the matrix of constant vectors. In addition, the determinant of the coefficient matrix is returned as a separate parameter.

## Running the Alternative Version of the Gauss-Jordan Elimination Program

Type the program shown in Listing 4.5, give it the name SOLVGV, and run it. As before, you are asked for the number of equations. Next

```
PROGRAM Solvgv;
{ Turbo Pascal program to perform simultaneous
{ solution by Gauss-Jordan elimination
  with multiple constant vectors }

CONST
  Maxr = 7;
  Maxc = 7;

TYPE
  Ary2s = ARRAY[1..Maxr, 1..Maxc] OF Real;

VAR
      A, Y: Ary2s;
  N, Nvec: Integer;
    Error: Boolean;
   Determ: Real;

PROCEDURE Get_Data(VAR A: Ary2s;
                   VAR Y: Ary2s;
                VAR N, Nvec: Integer);
{ Get values for N, Nvec, and arrays A, Y }

VAR
  I, J: Integer;

BEGIN
  WriteLn;
  REPEAT
    Write(' How many equations? ');
    ReadLn(N)
  UNTIL N < Maxr;
  IF N > 1 THEN
    BEGIN
      Write(' How many constant vectors? ');
      ReadLn(Nvec);
      FOR I := 1 TO N DO
        BEGIN
          FOR J := 1 TO N DO
            BEGIN
              Write(J:3, ': ');
              Read(A[I,J])
            END;
          IF Nvec > 0 THEN
            BEGIN
              FOR J := 1 TO Nvec DO
                BEGIN
                  Write(', C: ');
                  Read(Y[I,J])
                END;

              WriteLn;
            END
          ELSE {matrix determinant only}
              WriteLn;
        END; { I loop }
```

**Listing 4.5:** Solution of simultaneous equations and matrix inversion by the Gauss-Jordan elimination method (multiple constant vectors may be entered).

```
          WriteLn;
          Write('            Matrix');
          IF Nvec > 0 THEN
            Write('            Constants');
          WriteLn;
          FOR I:= 1 TO N  DO
            BEGIN
              FOR J:= 1 TO N DO
                Write(A[I,J]:7:4, '  ');
              FOR J := 1 TO Nvec DO
                Write(' :', Y[I,J]:7:4);
              WriteLn
            END; { I loop }
          WriteLn
      END  { if N>1 }
END; { procedure Get_Data }

PROCEDURE Write_Data;
{ print out the answers }

VAR
  I, J: Integer;

BEGIN
  IF Nvec > 0 THEN
    BEGIN
      WriteLn(' Solution');
      FOR I := 1 TO N DO
        BEGIN
          FOR J := 1 TO Nvec DO
            Write(Y[I,J]:9:5);
          WriteLn
        END
    END { if }
  ELSE
    BEGIN
      WriteLn('      Inverse');
      FOR I := 1 TO N DO
        BEGIN
          FOR J := 1 TO N DO
            Write('   ',A[I,J]:10);
          WriteLn
        END;
      WriteLn;
      Write(' Determinant is ', Determ:12)
    END; { ELSE }
  WriteLn
END; { Write_Data }
```

**Listing 4.5:** Solution of simultaneous equations and matrix inversion by the Gauss-Jordan elimination method (multiple constant vectors may be entered) (continued).

```
PROCEDURE Gaussjv
   (VAR B : Ary2s;  { square matrix of coefficients }
    VAR W : Ary2s;  { constant vector matrix }
    VAR Determ: Real;  { the determinant }
    Ncol      : Integer;  { order of matrix }
    Nv        : Integer;  { number of constants }
     VAR Error: Boolean); { True of matrix singular }

 {       Gauss Jordan matrix inversion and solution
   B(N,N) coefficient matrix, becomes inverse
   W(N,M) constant vector(s) become solution vector
   Determ is determinant
   Error = 1 if singular
   Index(N,3)
   Nv is number of constant vectors }

VAR
   Index: ARRAY[1..Maxc, 1..3] OF Integer;
   I, J, K, L, Irow, Icol, N, L1: Integer;
   Pivot, Hold, Sum, T, Ab, Big: Real;

PROCEDURE Swap(VAR A, B: Real);

VAR
   Hold: Real;

BEGIN  { Swap }
   Hold := A;
   A := B;
   B := Hold
END; { procedure Swap }

BEGIN      { Gauss-Jordan main program }
   Error := False;
   N := Ncol;
   FOR I := 1 TO N DO
      Index[I, 3] := 0;
   Determ := 1.0;
   FOR I := 1 TO N DO
     BEGIN
       { search for largest element }
       Big := 0.0;
       FOR J := 1 TO N DO
         BEGIN
           IF Index[J, 3] <> 1 THEN
             BEGIN
               FOR K := 1 TO N DO
                 BEGIN
                   IF Index[K, 3] > 1 THEN
                     BEGIN
                       WriteLn(' ERROR: matrix singular');
                       Error := True;
                       Exit  {procedure Gaussj }
                     END;
```

**Listing 4.5:** Solution of simultaneous equations and matrix inversion by the Gauss-Jordan elimination method (multiple constant vectors may be entered) (continued).

```
                        IF Index[K, 3] < 1 THEN
                          IF Abs(B[J, K]) > Big THEN
                            BEGIN
                              Irow := J;
                              Icol := K;
                              Big := Abs(B[J, K])
                            END
                      END { K loop }
              END
          END; { J loop }
      Index[Icol, 3] := Index[Icol, 3] + 1;
      Index[I, 1] := Irow;
      Index[I, 2] := Icol;

{ interchange rows to put pivot on diagonal }
IF Irow <> Icol THEN
  BEGIN
    Determ := - Determ;
    FOR L := 1 TO N DO
      Swap(B[Irow, L], B[Icol, L]);
    IF Nv > 0 THEN
      FOR L := 1 TO Nv DO
        Swap(W[Irow, L], W[Icol, L])
  END; { if Irow <> Icol }

    { divide pivot row by pivot column }
    Pivot := B[Icol, Icol];
    Determ := Determ * Pivot;
    B[Icol, Icol] := 1.0;
    FOR L := 1 TO N DO
      B[Icol, L] := B[Icol, L] / Pivot;
    IF Nv > 0 THEN
      FOR L := 1 TO Nv DO
        W[Icol, L] := W[Icol, L] / Pivot;
    {  reduce nonpivot rows }
    FOR L1 := 1 TO N DO
      BEGIN
        IF L1 <> Icol THEN
          BEGIN
            T := B[L1, Icol];
            B[L1, Icol] := 0.0;
            FOR L := 1 TO N DO
              B[L1, L] := B[L1, L] - B[Icol, L] * T;
            IF Nv > 0 THEN
              FOR L := 1 TO Nv DO
                W[L1, L] := W[L1, L] - W[Icol, L] * T;
          END   { IF L1 <> Icol }
      END
  END; { I loop }
```

**Listing 4.5:** Solution of simultaneous equations and matrix inversion by the Gauss-Jordan elimination method (multiple constant vectors may be entered) (continued).

```
         { interchange columns }
         FOR I := 1 TO N DO
           BEGIN
             L := N - I + 1;
             IF Index[L, 1] <> Index[L, 2] THEN
               BEGIN
                 Irow := Index[L, 1];
                 Icol := Index[L, 2];
                 FOR K := 1 TO N DO
                   Swap(B[K, Irow], B[K, Icol])
               END { if Index }
           END; { I loop }
         FOR K := 1 TO N DO
           IF Index[K, 3] <> 1 THEN
             BEGIN
               WriteLn(' ERROR: matrix singular');
               Error := True;
               Exit  {procedure Gaussj }
             END;
       END; { procedure Gaussjv }

       BEGIN   { main program }
         WriteLn;
         WriteLn(' Simultaneous solution by Gauss-Jordan');
         WriteLn
           (' Multiple constant vectors, or matrix inverse');
         REPEAT
           Get_Data(A, Y, N, Nvec);
           IF N > 1 THEN
             BEGIN
               Gaussjv(A, Y, Determ, N, Nvec, Error);
               IF NOT Error THEN  Write_Data
             END
         UNTIL N < 2
       END.
```

**Listing 4.5:** Solution of simultaneous equations and matrix inversion by the Gauss-Jordan elimination method (multiple constant vectors may be entered) (continued).

the program displays a question about the number of constant vectors. If you answer this question with the value of unity, then the program will behave exactly as before. However, it will display the solution vector vertically rather than horizontally.

With this new version, the number of constant vectors may be greater than one. In this case, you enter the constants for each equation on the same line as the corresponding coefficients. For example, if we solve the three electrical circuit configurations presented in the previous section collectively with this program, then the first line of data will be

13 −8 −3   20 20 −20

Solve all three of these circuits at the same time, and verify that the answers are correct.

This new version has another option. The number of constant vectors may be zero. In this case, you enter only a matrix of coefficients. The program will then display the inverse of the coefficient matrix, as well as the determinant of the coefficient matrix. Let us try this feature.

Enter the value 5 for the number of equations and the value 0 for the number of constant vectors. Then enter the following matrix:

$$\begin{bmatrix} 1. & 0.5 & 0.333333 & 0.25 & 0.2 \\ 0.5 & 0.333333 & 0.25 & 0.2 & 0.166667 \\ 0.333333 & 0.25 & 0.2 & 0.166667 & 0.142857 \\ 0.25 & 0.2 & 0.166667 & 0.142857 & 0.125 \\ 0.2 & 0.166667 & 0.142857 & 0.125 & 0.111111 \end{bmatrix}$$

We will look at this matrix in more detail in the next section. For now, we only have to know that it is nearly singular; that is, the determinant is very nearly zero. The result, given in Figure 4.3, shows that the determinant is about $10^{-12}$. As usual, you can terminate the program by pressing Ctrl-C.

In the following section, we will discuss the Hilbert matrix as an example of ill conditioning. To explore such matrices, we will write a program that uses our first version of the Gauss-Jordan procedure (Listing 4.4). We will then use the program to solve a series of progressively more problematic Hilbert matrices.

```
          Matrix
 1.0000  0.5000  0.3333   0.2500   0.2000
 0.5000  0.3333  0.2500   0.2000   0.1667
 0.3333  0.2500  0.2000   0.1667   0.1429
 0.2500  0.2000  0.1667   0.1429   0.1250
 0.2000  0.1667  0.1429   0.1250   0.1111

         Inverse
 2.542E+001  -3.08E+002  1.085E+003  -1.45E+003  6.561E+002
-3.08E+002   4.951E+003  -1.96E+004  2.788E+004  -1.31E+004
 1.084E+003  -1.96E+004  8.224E+004  -1.22E+005  5.884E+004
-1.45E+003   2.787E+004  -1.22E+005  1.858E+005  -9.15E+004
 6.555E+002  -1.31E+004  5.883E+004  -9.14E+004  4.569E+004

Determinant is 3.61625E-012
```

**Figure 4.3:** Matrix inverse of a Hilbert matrix by the Gauss-Jordan elimination method.

# Ill-Conditioned Equations

A singular matrix has a determinant of zero. The corresponding set of equations has either no solutions or many solutions. An *ill-conditioned matrix* is one that is *nearly* singular. It produces incorrect answers. A small change in the input data can cause great changes in the answer. A two-dimensional analogue of ill conditioning occurs with two nearly parallel lines. The point of intersection is the desired solution but, the closer the two lines come to being parallel, the more difficult it is to determine their actual point of intersection.

Various tests for ill conditioning have been suggested. One of these is to compare the values of the inverted matrix to those of the original matrix. If there are differences of several orders of magnitude, then it is likely that ill conditioning is present. Another test is to take the inverse of the inverse of the coefficient matrix and compare the result to the original matrix. Of course, the two should be the same. This will test the inversion algorithm and the computer arithmetic at the same time. Yet another test is to compare the magnitudes of values along the major diagonal. They should not be too far apart.

## The Hilbert Matrix

The Hilbert matrix is an example of ill conditioning. This symmetric matrix begins with unity in the upper-left corner. The remaining values get smaller and smaller as we go down a column or across a row, according to the pattern

$$
\begin{bmatrix}
1 & \frac{1}{2} & \frac{1}{3} & \frac{1}{4} & \cdots\cdots & 1/n \\
\frac{1}{2} & \frac{1}{3} & \frac{1}{4} & \frac{1}{5} & \cdots & 1/(n+1) \\
\frac{1}{3} & \frac{1}{4} & \frac{1}{5} & \frac{1}{6} & \cdots & 1/(n+2) \\
\frac{1}{4} & \frac{1}{5} & \frac{1}{6} & \frac{1}{7} & \cdots & 1/(n+3) \\
\cdots & & \cdots & & & \cdots \\
1/n & \cdots\cdots\cdots & & & & 1/(2n-1)
\end{bmatrix}
$$

The Hilbert matrix can be used to produce a set of progressively ill-conditioned equations. Consider, for example, the equations

$$
x_1 + \frac{x_2}{2} + \frac{x_3}{3} + \frac{x_4}{4} + \frac{x_5}{5} = 1 + \tfrac{1}{2} + \tfrac{1}{3} + \tfrac{1}{4} + \tfrac{1}{5}
$$

$$\frac{x_1}{2} + \frac{x_2}{3} + \frac{x_3}{4} + \frac{x_4}{5} + \frac{x_5}{6} = \tfrac{1}{2} + \tfrac{1}{3} + \tfrac{1}{4} + \tfrac{1}{5} + \tfrac{1}{6}$$

$$\frac{x_1}{3} + \frac{x_2}{4} + \frac{x_3}{5} + \frac{x_4}{6} + \frac{x_5}{7} = \tfrac{1}{3} + \tfrac{1}{4} + \tfrac{1}{5} + \tfrac{1}{6} + \tfrac{1}{7}$$

$$\frac{x_1}{4} + \frac{x_2}{5} + \frac{x_3}{6} + \frac{x_4}{7} + \frac{x_5}{8} = \tfrac{1}{4} + \tfrac{1}{5} + \tfrac{1}{6} + \tfrac{1}{7} + \tfrac{1}{8}$$

$$\frac{x_1}{5} + \frac{x_2}{6} + \frac{x_3}{7} + \frac{x_4}{8} + \frac{x_5}{9} = \tfrac{1}{5} + \tfrac{1}{6} + \tfrac{1}{7} + \tfrac{1}{8} + \tfrac{1}{9}$$

First, the fractions 1/3, 1/6, and 1/7 cannot be represented exactly. Therefore, there will be roundoff errors at the very beginning of the calculation. Second, the inverse of the coefficient matrix is exactly

$$\begin{bmatrix} 25 & -300 & 1050 & -1400 & 630 \\ -300 & 4800 & -18900 & 26880 & -12600 \\ 1050 & -18900 & 79380 & -117600 & 56700 \\ -1400 & 26880 & -117600 & 179200 & -88200 \\ 630 & -12600 & 56700 & -88200 & 44100 \end{bmatrix}$$

Some of the elements of the inverse matrix are orders of magnitude larger than elements of the original matrix. Furthermore, the determinant of the coefficient matrix is nearly zero.

Since each element of the constant vector is the sum of the matrix elements in the corresponding row, the exact solution is

$$[1 \quad 1 \quad 1 \quad 1 \quad 1]$$

But, because of the ill conditioning, the solution will be a little different. The usual Pascal, Fortran, or BASIC compiler gives a solution like this:

$$[1.0001 \quad 0.99816 \quad 1.00778 \quad 0.9884 \quad 1.0056]$$

However, Turbo Pascal is much more accurate, as we will see shortly.

Now let us look at a program that uses Hilbert matrices to study the effect of ill conditioning.

The program shown in Listing 4.6 will generate a set of Hilbert matrices and constant vectors corresponding to a solution of

$$[1 \quad 1 \quad 1 \quad \ldots \quad 1]$$

```
PROGRAM Solvit;
{ N x N inverse Hilbert matrix }
{ solution is 1 1 1 1 1 }

{ Turbo Pascal program to perform simultaneous
{ solution by Gauss-Jordan elimination }

CONST
  Maxr = 10;
  Maxc = 10;

TYPE
  Ary   = ARRAY[1..Maxr] OF Real;
  Arys  = ARRAY[1..Maxc] OF Real;
  Ary2s = ARRAY[1..Maxr, 1..Maxc] OF Real;

VAR
      Y, Coef: Arys;
         A, B: Ary2s;
  N, M, I, J: Integer;
        Error: Boolean;

PROCEDURE Get_Data(VAR A: Ary2s;
                   VAR Y: Arys;
                   VAR N, M: Integer);
{ Setup N-by-N hilbert matrix }

VAR
  I, J: Integer;

BEGIN
  FOR I := 1 TO N-1 DO
    BEGIN
      A[N,I] := 1.0/(N + I - 1);
      A[I,N] := A[N,I]
    END;
  A[N,N] := 1.0/(2*N -1);
  FOR I := 1 TO N DO
    BEGIN
      Y[I] := 0.0;
      FOR J := 1 TO N DO
        Y[I] := Y[I] + A[I,J]
    END;
  WriteLn;
  IF N < 7 THEN
    BEGIN
      FOR I:= 1 TO N  DO
        BEGIN
          FOR J:= 1 TO M DO
            Write(A[I,J]:7:5, '  ');
          WriteLn(' : ', Y[I]:7:5)
        END;
      WriteLn
    END  { if N<7 }
END; { procedure Get_Data }
```

**Listing 4.6:** Solution of a set of ill-conditioned equations.

```
PROCEDURE Write_Data;
{ print out the answers }

VAR
  I: Integer;

BEGIN
  WriteLn(' Solution for', M:3, ' equations');
  FOR I := 1 TO M DO
    Write(Coef[I]:13:9);
  WriteLn;
END; { Write_Data }

{$I GAUSSJ.PAS }  {Listing 4.4}

BEGIN  { main program }
  A[1,1] := 1.0;
  N := 2;
  M := N;
  REPEAT
    Get_Data(A, Y, N, M);
    FOR I := 1 TO N DO
      FOR J := 1 TO N DO
        B[I,J] := A[I,J]; { Setup work array }
    Gaussj(B, Y, Coef, N, Error);
    IF NOT Error THEN Write_Data;
    N := N+1;
    M := N
  UNTIL N > Maxr
END.
```

**Listing 4.6:** Solution of a set of ill-conditioned equations (continued).

## *Running the Hilbert Matrix Program*

You will not have to type the program given in Listing 4.6 because it is included on your Turbo Pascal disk. You gave it the name HILB.PAS earlier in this chapter. The Gauss-Jordan procedure is also included. When you start this program, it will run automatically. The program begins by solving the two equations

$$\begin{bmatrix} 1 & 0.5 \\ 0.5 & 0.3333 \end{bmatrix} \begin{bmatrix} 1.5 \\ 0.833 \end{bmatrix}$$

The determinant for this matrix is 0.0833, therefore this matrix is not ill conditioned. The solution vector is

$$\begin{bmatrix} 1 & 1 \end{bmatrix}$$

as expected. The program then continues with three, four, five, six, and seven equations.

Run the Hilbert matrix program and compare your results with Figure 4.4 for Turbo Pascal or Figure 4.5 for Turbo-87. Many compilers employ 32 bits of precision for floating-point operations. Then a 5-by-5 Hilbert matrix will show roundoff errors. However, you can see that both versions of Turbo Pascal use higher precision.

```
     1.00000   0.50000   0.33333   0.25000   0.20000   0.16667   :  2.45000
     0.50000   0.33333   0.25000   0.20000   0.16667   0.14286   :  1.59286
     0.33333   0.25000   0.20000   0.16667   0.14286   0.12500   :  1.21786
     0.25000   0.20000   0.16667   0.14286   0.12500   0.11111   :  0.99563
     0.20000   0.16667   0.14286   0.12500   0.11111   0.10000   :  0.84563
     0.16667   0.14286   0.12500   0.11111   0.10000   0.09091   :  0.73654

     Solution for   6 equations
       1.000000002    0.999999968    1.000000108    0.999999911    0.999999941
       1.000000070

     Solution for   7 equations
       0.999999991    1.000000420    0.999995642    1.000017765    0.999966353
       1.000029770    0.999990049

     Solution for   8 equations
       0.999999900    1.000005159    0.999934411    1.000347899    0.999077917
       1.001288431    0.999092482    1.000253859
```

**Figure 4.4:** Turbo Pascal solution to the Hilbert matrix program.

```
     Solution for   6 equations
       1.000000000    1.000000000    1.000000000    1.000000000    1.000000000
       1.000000000

     Solution for   7 equations
       1.000000000    1.000000000    0.999999997    1.000000012    0.999999978
       1.000000019    0.999999994

     Solution for   8 equations
       1.000000000    1.000000002    0.999999973    1.000000144    0.999999610
       1.000000554    0.999999605    1.000000112

     Solution for   9 equations
       1.000000000    1.000000016    0.999999730    1.000001919    0.999992987
       1.000014275    0.999983648    1.000009856    0.999997569

     Solution for  10 equations
       0.999999999    1.000000109    0.999997673    1.000021139    0.999899238
       1.000276832    0.999546040    1.000438494    0.999769892    1.000050585
```

**Figure 4.5:** Turbo-87 Pascal solution to the Hilbert matrix program.

Thus, this program can be used to test the significance of floating-point operations for any compiler.

Let us consider a variation of our equation-solving technique that is useful for working with large numbers of equations.

# *Reading Coefficients from a Disk File*

We have been solving simultaneous equations by entering the coefficients and constants from the keyboard. This is useful for solving small numbers of equations, but it is very tedious when you are working with larger numbers of equations. There is a better way to enter the values. For the next version of our program, we will create a disk file containing the coefficient matrix and constant vector. Then we will run a program that reads the data from the disk file.

## *Running the Disk-File Program*

Create a file named FITFILE.PAS containing the program shown in Listing 4.7. As before, the Gauss-Jordan procedure is required.

```
PROGRAM Fitfile;
{ Turbo Pascal program to perform simultaneous
{ solution by Gauss-Jordan elimination }
{ Read data from disk file }

CONST
   Maxr = 8;
   Maxc = 8;

TYPE
   Ary   = ARRAY[1..Maxr] OF Real;
   Arys  = ARRAY[1..Maxc] OF Real;
   Ary2s = ARRAY[1..Maxr, 1..Maxc] OF Real;

VAR
       Y, Coef: Arys;
          A, B: Ary2s;
   N, M, I, J: Integer;
         Error: Boolean;
      Filename: STRING[14];
        Filvar: Text;
```

**Listing 4.7:** Reading the data from a disk file.

```
PROCEDURE Get_Data(VAR A: Ary2s;
                   VAR Y: Arys;
                 VAR N, M: Integer);
{ Get values for N and arrays A, Y }

VAR
  I, J: Integer;

BEGIN { Get_Data }
   WriteLn;
   REPEAT
      Write(' Name of file: ');
      ReadLn(Filename);
      Assign(Filvar,Filename);
      {$I-}  {Turn off Error checking}
      Reset(Filvar);
      {$I+}  {Turn it back on}
   UNTIL IOresult = 0;
   ReadLn(Filvar, N);  { number of equations }
   M := N;
   WriteLn;
   WriteLn
    (' Simultaneous solution by Gauss-Jordan elimination');
   WriteLn(' ',N,' equations ');

   IF N > 1 THEN
     BEGIN
       FOR I := 1 TO N DO
         BEGIN
           FOR J := 1 TO N DO
               {Read(A[I,J])}
               Read(Filvar, A[I,J]);
           ReadLn(Filvar, Y[I])   { clear line }
         END;

       FOR I:= 1 TO N  DO
         BEGIN
           FOR J:= 1 TO M DO
             Write(A[I,J]:7:4, '  ');
           WriteLn(' : ', Y[I]:7:4)
         END;
       WriteLn
     END  { if N>1 }
END; { procedure Get_Data }

PROCEDURE Write_Data;
{ print out the answers }

VAR
  I: Integer;

BEGIN
  FOR I := 1 TO M DO
    Write(Coef[I]:9:5);
  WriteLn
END; { Write_Data }
```

**Listing 4.7:** Reading the data from a disk file (continued).

```
{$I GAUSSJ.PAS} {Listing 4.4}

BEGIN  { main program }
  WriteLn;
   REPEAT
    Get_Data(A, Y, N, M);
   IF N > 1 THEN
      BEGIN
        FOR I := 1 TO N DO
          FOR J := 1 TO N DO
            B[I,J] := A[I,J]; { Setup work array }
          Gaussj(B, Y, Coef, N, Error);
          IF NOT Error THEN Write_Data
      END
   UNTIL N < 2
END.
```

**Listing 4.7:** Reading the data from a disk file (continued).

Procedure Get_Data asks you for the disk file name. Enter the name in the usual DOS fashion (type the first name, a period, and then the extension as needed). If the name you give does not exist, the program repeats the request.

Create a disk file named FIT3.TXT that contains the data

```
 3
13   -8   -3   20
-8   10   -1   -5
-3   -1   11    0
```

which we used earlier in this chapter. (You can use the Turbo Pascal editor.) The first line of this data file has only the value 3, the number of simultaneous equations to be solved. The next three lines give the coefficients and then the constant, one equation to each line. The spacing is not important; just place a space or two between each value.

Make a second data file named FIT5.TXT, and type the lines

```
 5
 1    1    1    1    1   15
 1    1   -1   -1    1    1
 2   -1    2   -1    2   12
 5    4    3    2    1   35
-1    2   -3    4   -5  -15
```

corresponding to five simultaneous equations. Run the FITFILE program, and it will ask you for the name of the file containing the data. Answer

    FIT3.TXT

The program displays the input data and then the solution. Then it cycles so you can solve another set of equations. Give the name

    FIT5.TXT

to solve the second set of equations. The solution to this second set is

    1.00000  2.00000  3.00000  4.00000  5.00000

The advantage of entering the data into a disk file is apparent when there are many numbers (there were 31 in this last example). Then, if you discover an error in your data, you can easily edit the data file to make the correction. With our previous programs you have to reenter all the numbers if there is an error.

Our program does not analyze the data file. Therefore, a useful addition would be a routine to check the data for consistency. An error message could be displayed if the data file is inconsistent, which could be because it has too few equations or missing coefficients.

In the next section, we will see another variation of the Gauss-Jordan procedure. We will consider a set in which the number of equations is greater than the number of variables in each equation, and we will present a program to compute the best-fit solution.


## A Simultaneous Best Fit

The previous programs in this chapter produce an *exact* fit to a set of linear equations. In these examples, the number of unknowns equals the number of independent equations. If, however, there are more unknowns then there are equations, no unique solution is possible. This corresponds to a coefficient matrix having more columns than rows.

There is another case we might consider. Suppose that we want to determine the value of $M$ unknowns by an experimental procedure. If the number of independent measurements is the same, then we can find an exact solution. On the other hand, suppose that the number of independent measurements, $N$, is greater than the number of unknowns, $M$. Then it is possible to calculate a *best-fit* value for the $M$ unknowns.

As a two-dimensional analogue, consider the three equations

$$x + y = 3$$
$$x \qquad = 1$$
$$\qquad y = 1$$

These three lines do not intersect at the same point. Rather, each pair of lines defines a different point on the x-y plane. These points describe a right triangle that has corners at the positions $(1, 1)$, $(2, 1)$, and $(1, 2)$, as shown in Figure 4.6. The best choice for the "intersection" of these three lines is the location of the centroid of the triangle. Since this point is located at one-third of the distance from the base to the apex, then the best-fit solution to these three lines is

$$x = 1.3333, \qquad y = 1.3333$$

Now that we have illustrated a best-fit solution, we will look at the program that will find such a solution.

## *Programming the Best-Fit Solution*

Make a copy of the program shown in Listing 4.3, and alter it to look like Listing 4.8. Procedure Square developed in the previous chapter is

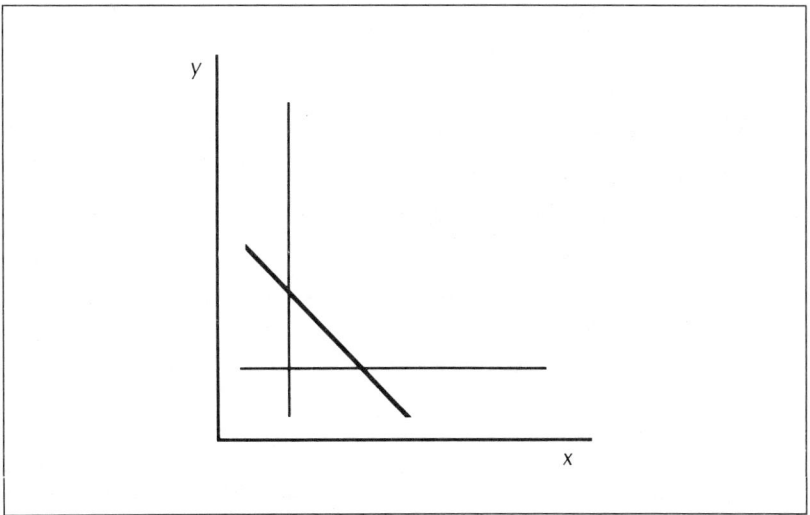

**Figure 4.6:** Two-dimensional best fit.

```
PROGRAM Solvgj;
{ Turbo Pascal program to perform simultaneous
{ solution by Gauss-Jordan elimination }
{ there may be more equations than unknowns }

CONST
  Maxr = 8;
  Maxc = 8;

TYPE
  Ary   = ARRAY[1..Maxr] OF Real;
  Arys  = ARRAY[1..Maxc] OF Real;
  Ary2s = ARRAY[1..Maxr, 1..Maxc] OF Real;
  Ary2  = Ary2s; { for Square }

VAR
          Y: Ary;
   Coef, Yy: Arys;
      A, B: Ary2s;
  N, M, I, J: Integer;
      Error: Boolean;

PROCEDURE Get_Data(VAR A: Ary2s;
                   VAR Y: Ary;
                 VAR N, M: Integer);
{ Get values for N and arrays A, Y }

VAR
  I, J: Integer;

BEGIN
  WriteLn;
  REPEAT
    Write(' How many unknowns? ');
    ReadLn(M)
  UNTIL M < Maxc;
  IF M > 1 THEN
    BEGIN
      REPEAT
        Write(' How many equations? ');
        ReadLn(N)
      UNTIL N >= M;
      FOR I := 1 TO N DO
        BEGIN
          WriteLn(' Equation', I:3);
          FOR J := 1 TO M DO
            BEGIN
              Write(J:3, ': ');
              Read(A[I,J])
            END;
          Write(', C: ');
          ReadLn(Y[I])   { clear line }
        END; { I loop }
```

**Listing 4.8:** The best fit to a set of linear equations.

```
            WriteLn;
            FOR I:= 1 TO N  DO
               BEGIN
                  FOR J:= 1 TO M DO
                     Write(A[I,J]:7:4, '  ');
                  WriteLn(' : ', Y[I]:7:4)
               END;
            WriteLn
         END  { if N>1 }
END; { procedure Get_Data }

PROCEDURE Write_Data;

{ print out the answers }

VAR
   I: Integer;

BEGIN
   FOR I := 1 TO M DO
      Write(Coef[I]:9:5);
   WriteLn
END; { Write_Data }

{$I SQUARE.PAS } {Listing 3.2}
{$I GAUSSJ.PAS } {Listing 4.4}

BEGIN  { main program }
   WriteLn;
   WriteLn(' Best fit to simultaneous equations');
   WriteLn(' By Gauss-Jordan');
   REPEAT
      Get_Data(A, Y, N, M);
      IF M > 1 THEN
         BEGIN
            Square(A, Y, B, Yy, N, M);
            Gaussj(B, Yy, Coef, M, Error);
            IF NOT Error THEN Write_Data
         END
   UNTIL M < 2
END.
```

**Listing 4.8:** The best fit to a set of linear equations (continued).

needed. It is referenced by an INCLUDE directive. Remember, this procedure converts the rectangular array of coefficients and the constant vector into the square array and vector needed by procedure Gaussj, which is also referenced through an INCLUDE directive.

The technique used in this program is essentially the same as that used for least-squares curve fitting. Since this topic is covered in the next chapter, we will not discuss it here.

## *Running the Best-Fit Program*

When you run the best-fit program, it begins a little differently than the others that we have developed. It requests the number of unknowns first, then it asks for the number of equations. If these are the same value, then the simultaneous solution is carried out as before. However, if there are more equations than unknowns, then the best fit is returned. The three equations that we discussed earlier in this section have only two unknowns; the matrix and vector are

$$\begin{bmatrix} 1 & 1 \\ 1 & 0 \\ 0 & 1 \end{bmatrix} \begin{bmatrix} 3 \\ 1 \\ 1 \end{bmatrix}$$

Enter these data into the program, and verify that the result is

x = 1.3333, y = 1.3333

as shown in Figure 4.7.

As another example, consider the electrical circuit presented earlier in the chapter. We originally derived three loop-current equations that were solved simultaneously. Suppose, however, that the source voltages were determined experimentally. If the value of the left source was found to be 19 volts and the value of the right source was measured as

```
Best fit to simultaneous equations
By Gauss-Jordan

How many unknowns? 2
How many equations? 3
Equation  1
  1: 1  2: 1, C: 3
Equation  2
  1: 1  2: 0, C: 1
Equation  3
  1: 0  2: 1, C: 1

1.0000    1.0000    :   3.0000
1.0000    0.0000    :   1.0000
0.0000    1.0000    :   1.0000

  1.33333   1.33333
```

**Figure 4.7:** Best-fit simultaneous solution.

$-5.1$ volts, then the three loop equations would be

$$\begin{bmatrix} 13 & -8 & -3 \\ -8 & 10 & -1 \\ -3 & -1 & 11 \end{bmatrix} \begin{bmatrix} 19 \\ -5.1 \\ 0 \end{bmatrix}$$

In addition, suppose that the voltage across the horizontal, 1-ohm resistor was measured to be 1.1 volts. The branch current flowing through this resistor is loop current 2 minus loop current 3. Consequently, we can now write an additional independent equation, corresponding to an additional row in our matrix. The four equations are

$$\begin{bmatrix} 13 & -8 & -3 \\ -8 & 10 & -1 \\ -3 & -1 & 11 \\ 0 & 1 & -1 \end{bmatrix} \begin{bmatrix} 19 \\ -5.1 \\ 0 \\ 1.1 \end{bmatrix}$$

Enter these four equations, with their three unknowns, into the new program. Compare the resulting best-fit solution

$$I_1 = 2.8 \text{ amps}, I_2 = 1.8 \text{ amps}, I_3 = 0.94 \text{ amps}$$

to the original solution. This program can be terminated by entering zero for the number of equations or by pressing Ctrl-C.

Next we will devise a method—and a Turbo Pascal program—for solving simultaneous equations that have complex coefficients (that is, factors containing imaginary parts). To illustrate this problem, we will study a second, more complicated electrical example.

## *Equations with Complex Coefficients*

Simultaneous equations with complex coefficients are used in the analysis of electrical circuits. Complex numbers are not a standard type in Turbo Pascal, although they can be defined by the user. In fact, in the second edition of their *Pascal User Manual and Report*, Kathleen Jensen and Niklaus Wirth show how to generate complex numbers using record types. Nevertheless, it is rather easy to solve a set of complex simultaneous equations by converting them into twice the number of equations with real coefficients. The new set can then be solved by one of the methods developed previously in this chapter.

### *Example: An Alternating-Current Electrical Circuit*

Consider the electrical circuit shown in Figure 4.8. This circuit is more complicated than the one illustrated in Figure 4.1 since it contains an alternating-current (AC) power source, an inductor, and a capacitor. The impedance function for a resistor is simply the resistance $R$. However, the impedance for inductors and capacitors is a function of the frequency. The impedance function for the inductor is $j\omega L$, where $j$ is the imaginary operator equal to the square root of $-1$, $\omega$ is the frequency of the AC source in radians per second, and $L$ is the self inductance in henries. The impedance function for the capacitor is $-j/\omega C$, where $C$ is the capacitance in farads.

For Figure 4.8, the impedance functions are shown next to the corresponding elements. The AC power source is 10 volts (RMS) with a frequency of $\omega$. The inductor has an impedance of $j5$ ohms, and the capacitor has an impedance of $-j4$ ohms.

We can find the branch currents for this circuit by using the two loop currents and the Kirchhoff voltage law. A clockwise summing around each loop gives

$$(6 + j5)I_1 - 6I_2 - 10 = 0 \qquad \text{(left loop)}$$

$$-6I_1 + (8 - j4)I_2 = 0 \qquad \text{(right loop)}$$

The corresponding linear equations are

$$\begin{bmatrix} (6 + j5) & (-6 + j0) \\ (-6 + j0) & (8 - j4) \end{bmatrix} \begin{bmatrix} (10 + j0) \\ (0 + j0) \end{bmatrix}$$

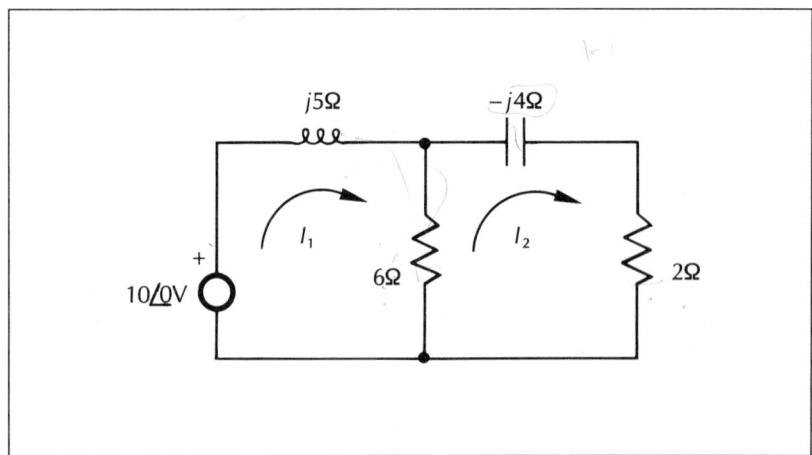

**Figure 4.8:** A network containing an AC supply.

These two equations cannot be directly solved with the programs given previously in this chapter because they contain complex coefficients.

Let us consider a general statement of the two loop equations. The electrical current and the impedance can both be expressed as complex numbers. Therefore, we can write

$$(AR_{11} + jAI_{11})(IR_1 + jII_1) + (AR_{21} + jAI_{21})(IR_2 + jII_2)$$
$$= (VR_1 + jVI_1)$$
$$(AR_{21} + jAI_{21})(IR_1 + jII_1) + (AR_{22} + jAI_{22})(IR_2 + jII_2)$$
$$= (VR_2 + jVI_2)$$

where the following symbols are used:

$$AR_{kl} = \text{real part of coefficient (impedance) } k,l$$
$$AI_{kl} = \text{imaginary part of coefficient } k,l$$
$$IR_l = \text{real part of current l}$$
$$II_l = \text{imaginary part of current l}$$
$$VR_k = \text{real part of voltage for equation } k$$
$$VI_k = \text{imaginary part of voltage for equation } k$$

Multiplication of the terms on the left of the above equations produces groups that alternately include the complex operator $j$.

$$(AR_{11}IR_1 - AI_{11}II_1) + j(AR_{11}II_1 + AI_{11}IR_1) + \ldots$$

But, if the complex expression on the left is to equal the complex expression on the right, then the real terms on the left must equal the real terms on the right. Similarly, the real coefficients of the imaginary terms on the left must equal the corresponding terms on the right. This approach creates a new set of $2N$ simultaneous equations that contain only real coefficients.

The first new equation is set equal to the real part of the first constant (voltage) term

$$AR_{11} - AI_{11} + AR_{12} - AI_{12} = VR_1$$

Notice that the complex conjugates of the original coefficients appear in the new first equation; that is, the original coefficients appear in order, but with alternating signs.

The second new equation is set equal to the imaginary part of the first constant (voltage) term

$$AI_{11} + AR_{11} + AI_{12} + AR_{12} = VI_1$$

This equation also contains all the coefficients for the first original equation. But now the real and imaginary parts are interchanged. Furthermore, the original signs are used.

The complete new equations can be summarized as

$$
\begin{bmatrix}
AR_{11} & -AI_{11} & AR_{12} & -AI_{12} & VR_1 \\
AI_{11} & AR_{11} & AI_{12} & AR_{12} & VI_1 \\
AR_{21} & -AI_{21} & AR_{22} & -AI_{22} & VR_2 \\
AI_{21} & AR_{21} & AI_{22} & AR_{22} & VI_2
\end{bmatrix}
$$

Substituting the values from Figure 4.8 gives

$$
\begin{bmatrix}
6 & -5 & -6 & 0 \\
5 & 6 & 0 & -6 \\
-6 & 0 & 8 & 4 \\
0 & -6 & -4 & 8
\end{bmatrix}
\begin{bmatrix}
10 \\
0 \\
0 \\
0
\end{bmatrix}
$$

Notice that each original coefficient appears twice in the new matrix. The solution vector for the new set of equations can readily be found by the methods of this chapter. The solution is

$$
[1.5 \quad -2.0 \quad 1.5 \quad -0.75]
$$

which corresponds to the loop currents

$$
I_1 = 1.5 - j2 \quad \text{amps} = 2.5 \quad \underline{/-53^\circ}
$$
$$
I_2 = 1.5 - j0.75 \text{ amps} = 1.67 \quad \underline{/-27^\circ}
$$

These results can be readily verified by calculating the voltages across each circuit element. For example, if the lower node is chosen to be zero volts, then the voltage of the upper node is equal to the voltage across the 6-ohm resistor:

$$
V = 6(I_1 - I_2) = 6(-j1.25) = -j7.5 \text{ volts}
$$

Similarly, the voltage across the inductor is

$$
V = j5(I_1) = 10 + j7.5 \text{ volts}
$$

A sum of the voltages around the left loop then gives

$$
(-j7.5) + (10 + j7.5) - (10) = 0
$$

A similar check can be made on the right loop.

Let us now look at a program that will handle these complex coefficients.

### *Programming Simultaneous Equations with Complex Coefficients*

The program given in Listing 4.9 simplifies the solution of simultaneous equations with complex coefficients. Each coefficient of the original $N$ equations is entered only once. Then the program converts the data into a $2N$-by-$2N$ matrix and a constant vector of length $2N$. Up to four complex equations can be solved simultaneously. This number can be increased by changing the values of Nrow and Ncol. (They must be twice the maximum number of equations.) We can use any of the

```
PROGRAM Solvec;
{ Turbo Pascal program to perform simultaneous
{ solution for complex coefficients }
{ by Gauss-Jordan elimination }

CONST
  Maxr = 8;
  Maxc = 8;

TYPE
  Ary   = ARRAY[1..Maxr] OF Real;
  Arys  = ARRAY[1..Maxc] OF Real;
  Ary2s = ARRAY[1..Maxr, 1..Maxc] OF Real;
  Aryc2 = ARRAY[1..Maxr, 1..Maxc, 1..2] OF Real;
  Aryc  = ARRAY[1..Maxr, 1..2] OF Real;

VAR
     Y, Coef: Arys;
        A, B: Ary2s;
  N, M, I, J: Integer;
       Error: Boolean;

PROCEDURE Get_Data(VAR A: Ary2s;
                   VAR Y: Arys;
                 VAR N, M: Integer);
{ Get complex values for N and arrays A, Y }

VAR
  C: Aryc2;
  V: Aryc;
  I, J, K, L: Integer;

PROCEDURE Show;
{ print original data }

VAR
  I, J, K: Integer;
```

**Listing 4.9:** Simultaneous solution of equations with complex coefficients

```
BEGIN { show }
  WriteLn;
  FOR I:= 1 TO N  DO
    BEGIN
      FOR J:= 1 TO M DO
        FOR K := 1 TO 2 DO
          Write(C[I,J,K]:7:4, '  ');
        WriteLn(' : ', V[I,1]:7:4,' : ',V[I,2]:7:4)
    END;
  N := 2 * N;
  M := N;
  WriteLn;
  FOR I:= 1 TO N  DO
    BEGIN
      FOR J:= 1 TO M DO
        Write(A[I,J]:7:4, '  ');
      WriteLn(' : ', Y[I]:9:5)
    END;
  WriteLn
END; { show }

BEGIN { procedure Get_Data }
  WriteLn;
  REPEAT
    Write(' How many equations? ');
    ReadLn(N);
    M := N
  UNTIL N < Maxr;
  IF N > 1 THEN
    BEGIN
      FOR I := 1 TO N DO
        BEGIN
          WriteLn('Equation', I:3);
          K := 0;
          L := 2 * I - 1;
          FOR J := 1 TO N DO
            BEGIN
              K := K + 1;
              Write(' Real ', J:3, ': ');
              Read(C[I,J, 1]); { real part }
              A[L,K] := C[I,J, 1];
              A[L+1,K+1] := C[I,J, 1];
              K := K + 1;
              Write(' Imag ', J:3, ': ');
              Read(C[I,J, 2]); {imaginary part}
              A[L,K] := -C[I,J, 2];
              A[L+1,K-1] := C[I,J, 2]
            END; { J loop }
          Write(' Real CONST: ');
          Read(V[I,1]); { real constant }
          Y[L] := V[I,1];
          Write(' Imag CONST: ');
          ReadLn(V[I,2]); { imaginary constant }
          Y[L+1] := V[I,2]
        END; { I loop }
```

**Listing 4.9:** Simultaneous solution of equations with complex coefficients (continued).

```
      Show { original data }
   END  { if N>1 }
END; { procedure Get_Data }

PROCEDURE Write_Data;
{ print out the answers }

VAR
    I, J: Integer;
  Re, Im: Real;

{$I ATAN.PAS } {Listing 1.3}

FUNCTION Mag(X, Y: Real): Real;
{ polar magnitude }
BEGIN
  Mag := Sqrt(Sqr(X) + Sqr(Y))
END; { function Mag }

BEGIN
  WriteLn
    ('      Real      Imaginary    Magnitude    Angle');
  FOR I := 1 TO M DIV 2 DO
    BEGIN
      J := 2*I -1;
      Re := Coef[J];
      Im := Coef[J+1];
      WriteLn(Re:11:5, Im:11:5,
        Mag(Re, Im):11:5,
        Atan(Re, Im):11:5)
    END; { FOR }
  WriteLn
END; { Write_Data }

{$I GAUSSJ.PAS } {Listing 4.4}

BEGIN  { main program }
  WriteLn;
  WriteLn
    (' Simultaneous solution with complex coefficients');
  WriteLn(' By Gauss-Jordan elimination');
  REPEAT
    Get_Data(A, Y, N, M);
    IF N > 1 THEN
      BEGIN
        FOR I := 1 TO N DO
          FOR J := 1 TO N DO
            B[I,J] := A[I,J]; { Setup work array }
        Gaussj(B, Y, Coef, N, Error);
        IF NOT Error THEN Write_Data
      END
  UNTIL N < 2
END.
```

**Listing 4.9:** Simultaneous solution of equations with complex coefficients (continued).

methods previously developed in this chapter to find the solution. However, let us continue with a copy of the Gauss-Jordan elimination program given in Listing 4.4.

### Running the Complex-Coefficient Program

Type the program shown in Listing 4.9, give it the name SOLVEC.PAS, and use it to solve the circuit shown in Figure 4.8. Procedure Gaussj and function Atan from Listing 1.3 are required. Run the program and enter the value of 2 for the number of complex equations. Then enter the coefficients for each equation in turn for the above problem. Give the real coefficient first, then the imaginary part next. Enter the constant vector in the same way; real part first, then imaginary part. The data for this problem are entered as

```
  6,   5,  −6,    0,   10,   0 (Equation 1)
 −6,   0,   8,   −4,    0,   0 (Equation 2)
```

The program displays the original input data, and then it shows the new $2N$-by-$2N$ matrix. Finally, it gives the solution in both rectangular and polar forms, as shown in Figure 4.9.

The polar magnitude is calculated in function Mag by finding the square root of the sum of the squares of the rectangular components. The phasor angle is determined in function Atan. This function appears

```
    Simultaneous solution with complex coefficients
    By Gauss-Jordan elimination

    How many equations? 2
  Equation  1
    Real 1: 6 Imag  1: 5 Real 2: −6 Imag 2: 0 Real CONST: 10 Imag CONST: 0
  Equation  2
    Real 1: −6 Imag 1: 0 Real 2: 8 Imag  2: −4 Real CONST: 0 Imag CONST: 0

    6.0000    5.0000   −6.0000    0.0000   : 10.0000 :   0.0000
   −6.0000    0.0000    8.0000   −4.0000   :  0.0000 :   0.0000

    6.0000   −5.0000   −6.0000    0.0000   : 10.00000
    5.0000    6.0000    0.0000   −6.0000   :  0.00000
   −6.0000    0.0000    8.0000    4.0000   :  0.00000
    0.0000   −6.0000   −4.0000    8.0000   :  0.00000

       Real    Imaginary   Magnitude   Angle
    1.50000    −2.00000     2.50000   −53.13010
    1.50000    −0.75000     1.67705   −26.56505
```

**Figure 4.9:** Simultaneous solution with complex coefficients.

to be more complicated than necessary. As we saw previously, function Atan computes the arc tangent for the first quadrant (0 to 90°). Then the angle is converted to the appropriate quadrant.

We have now seen several methods and variations that are adequate for solving small matrices (that is, solving small numbers of simultaneous equations). The last topic of this chapter will be an iterative method for solving large matrices.

# The Gauss-Seidel Iterative Method

The Gauss elimination and Gauss-Jordan elimination methods we considered previously are not suitable for solving very large matrices. More and more multiplications and subtractions are done as the number of equations increases. The resulting roundoff error can produce a meaningless solution.

The *Gauss-Seidel method* finds the solution to a set of equations by an iterative technique. An initial approximation is repeatedly refined until the result is acceptably close to the solution. Since each approximation depends only on the previous approximation, roundoff error does not accumulate. An added feature is that the equations do not have to be linear.

Consider the three loop-current equations derived from Figure 4.1:

$$13I_1 - 8I_2 - 3I_3 = 20$$
$$-8I_1 + 10I_2 - I_3 = -5$$
$$-3I_1 - I_2 + 11I_3 = 0$$

We can solve the first equation for $I_1$:

$$I_1 = \frac{20 + 8I_2 + 3I_3}{13}$$

in terms of the other two unknowns. Then if we chose first approximations of zero for $I_2$ and $I_3$, we obtain a value of 1.54 for $I_1$. The second equation is then solved for the second variable

$$I_2 = \frac{8I_1 + I_3 - 5}{10}$$

Substituting the current values of 1.54 for $I_1$ and 0 for $I_3$ produces a value of 0.73 for $I_2$. The third equation is similarly solved for the third variable:

$$I_3 = \frac{3I_1 + I_2}{11}$$

The current values of 1.54 for $I_1$ and 0.73 for $I_2$ give a value of 0.486 for $I_3$. The process is now repeated. The values of 0.73 for $I_2$ and 0.486 for $I_3$ are used to obtain a better value for $I_1$. After about 20 complete iterations, the values are correct to three significant figures. The following table gives some of the values in the sequence:

| $I_1$ | $I_2$ | $I_3$ |
|-------|-------|-------|
| 0     | 0     | 0     |
| 1.54  | 0.73  | 0.486 |
| 2.10  | 1.23  | 0.685 |
| 2.45  | 1.53  | 0.808 |
| 2.67  | 1.71  | 0.883 |

There are several potential problems with the Gauss-Seidel method. First, the process might not converge. That is, successive values may drift further and further from the correct solution. Consider, for example, the previous three equations. If they are written in reverse order, then we will derive the expressions

$$I_1 = (11I_3 - I_2)/3$$
$$I_2 = (13I_1 - 3I_3 - 20)/13$$
$$I_3 = 10I_2 - 8I_1$$

First approximations of zero are chosen, as before. However, the following values are obviously diverging:

| $I_1$ | $I_2$  | $I_3$ |
|-------|--------|-------|
| 0     | 0      | 0     |
| 0     | $-1.5$ | $-15$ |
| $-57$ | $-54$  | $-92$ |
| $-318$| $-299$ | $-440$|

The problem in this second example is that the largest values are not located on the major diagonal. The solution is to interchange rows to bring the largest element into the pivot position, as we did earlier in the chapter.

Finally, let us investigate a Turbo Pascal version of the Gauss-Seidel method. We will discuss the differences between relative and absolute criteria in IF ... THEN decisions and the meaning and use of point relaxation.

## *Programming the Gauss-Seidel Method*

Surprisingly, for the Gauss-Seidel method, the choice of a first approximation is not too important. An additional matter to be considered, however, is the criterion for convergence. The program shown in Listing 4.10 can be used to explore the Gauss-Seidel method for the solution of linear simultaneous equations. Most of the program can be copied from the Gauss elimination program given in Listing 4.2.

The pivot-interchange routine used in the Gauss elimination program is incorporated here for the same purpose. Rows are interchanged to place the largest element of each column on the major diagonal.

An *absolute*, rather than a relative, criterion is used to determine convergence:

    IF Abs(Nextc - Coef[j]) > Tol THEN ...

Normally, it is better to choose a relative criterion:

    IF Abs(1 - Coef[j] / Nextc) > Tol THEN ...

But, then we must ensure that Nextc will never be zero. One way to do this would be to use the form

    IF Abs(Nextc - Coef[j]) > Abs(Tol*Nextc) THEN ...

Sometimes the successive approximations jump about wildly. One feature of the program given in Listing 4.10 reduces this tendency. If two successive approximations differ in sign, then the step size is cut in half.

Another feature of the Gauss-Seidel program shown in Listing 4.10 is known as *point relaxation*. With this technique, each selected value is a function of the previous iteration (the calculated value) and a relaxation factor, Lambda. If Coef[j] is the previous value and Nextc is the calculated value, then the next value becomes

    Coef[j] := Lambda * Nextc + (1.0-Lambda) * Coef[j];

The value of Lambda can range from 0 to 2.

```pascal
PROGRAM Gausid;
{ Turbo Pascal program to perform
  simultaneous solution by Gauss-Seidel
  procedure Seid is included }

CONST
  Maxr = 8;
  Maxc = 8;

TYPE
  Ary   = ARRAY[1..Maxr] OF Real;
  Arys  = ARRAY[1..Maxc] OF Real;
  Ary2s = ARRAY[1..Maxr, 1..Maxc] OF Real;

VAR
      Y: Ary;
   Coef: Arys;
      A: Ary2s;
   N, M: Integer;
  Error: Boolean;

PROCEDURE Get_Data
     (VAR A: Ary2s;
      VAR Y: Ary;
   VAR N, M: Integer);
{ Get values for N and arrays A, Y }

VAR
  I, J: Integer;
BEGIN
  WriteLn;
  REPEAT
    Write(' How many equations? ');
    ReadLn(N)
  UNTIL N < Maxr;
  M := N;
  IF N > 1 THEN
    BEGIN
      FOR I := 1 TO N DO
        BEGIN
          WriteLn;
          WriteLn(' Equation', I: 3);
          FOR J := 1 TO N DO
            BEGIN
              Write(J:3, ': ');
              Read(A[I, J])
            END;
          Write(', C: ');
          Read(Y[I]);
        END;
      WriteLn;
      FOR I := 1 TO N DO
        BEGIN
          FOR J := 1 TO M DO
            Write(A[I, J]:7:4, '  ');
          WriteLn(' : ', Y[I]:7:4)
        END;
```

**Listing 4.10:** Solution of linear equations by the Gauss-Seidel method.

```
          WriteLn
      END  { if N>1 }
   ELSE IF N < 0 THEN N := - N;
   M := N
END; { procedure Get_Data }

PROCEDURE Write_Data;
      { print out the answers }
VAR
   I: Integer;

BEGIN
   FOR I := 1 TO M DO
     Write(Coef[I]:9:5);
   WriteLn
END; { Write_Data }

PROCEDURE Seid
           (A: Ary2s;
             Y: Ary;
      VAR Coef: Arys;
          Ncol: Integer;
      VAR Error: Boolean);
{ matrix solution by Gauss-Seidel }

CONST
   Tol = 1.0E-4;
   Max = 100;

VAR
   Done: Boolean;
   I, J, K, L, N: Integer;
   Nextc, Hold, Sum, Lambda, Ab, Big: Real;

BEGIN
   REPEAT
     Write(' Relaxation factor? ');
     ReadLn(Lambda)
   UNTIL (Lambda < 2.0) AND (Lambda > 0.0);
   Error := False;
   N := Ncol;
   FOR I := 1 TO N - 1 DO
     BEGIN
       Big := Abs(A[I, I]);
       L := I;
       FOR J := I + 1 TO N DO
         BEGIN
           { search for largest element }
           Ab := Abs(A[J, I]);
           IF Ab > Big THEN
             BEGIN
               Big := Ab;
               L := J
             END
```

**Listing 4.10:** Solution of linear equations by the Gauss-Seidel method (continued).

```
              END; { J loop }
          IF Big = 0.0 THEN
            Error := True
          ELSE
            BEGIN
              IF L <> I THEN
                BEGIN
                  { interchange rows to put }
                  { largest element on diagonal }
                  FOR J := 1 TO N DO
                    BEGIN
                      Hold := A[L, J];
                      A[L, J] := A[I, J];
                      A[I, J] := Hold
                    END;
                  Hold := Y[L];
                  Y[L] := Y[I];
                  Y[I] := Hold
                END   { if L <> I }
            END       { if Big }
        END;          { I loop }
      IF A[N, N] = 0.0 THEN
        Error := True
      ELSE
        BEGIN
          FOR I := 1 TO N DO
            Coef[I] := 0.0; { initial guess }
          I := 0;
          REPEAT
            I := I + 1;
            Done := True;
            FOR J := 1 TO N DO
              BEGIN
                Sum := Y[J];
                FOR K := 1 TO N DO
                  IF J <> K THEN
                    Sum := Sum - A[J, K] * Coef[K];
                Nextc := Sum / A[J, J];
                IF Abs(Nextc - Coef[J]) > Tol THEN
                  BEGIN
                    Done := False;
                    IF Nextc * Coef[J] < 0.0 THEN
                      Nextc := (Coef[J] + Nextc) * 0.5
                  END;
                Coef[J] :=
                  Lambda*Nextc + (1.0 - Lambda)*Coef[J];
                WriteLn(I:4, ',Coef(', J, ') =',Coef[J])
              END { J loop }
          UNTIL Done OR (I > Max)
        END; { IF A[N,N] = 0 }
      IF I > Max THEN Error := True;
      IF Error THEN WriteLn('ERROR: Matrix singular ')
    END; { Seid }
```

**Listing 4.10:** Solution of linear equations by the Gauss-Seidel method (continued).

```
BEGIN                 { main program }
   WriteLn;
   WriteLn(' Simultaneous solution by Gauss-Seidel');
   REPEAT
     Get_Data(A, Y, N, M);
     IF N > 1 THEN
       BEGIN
         Seid(A, Y, Coef, N, Error);
         IF NOT Error THEN Write_Data
       END
   UNTIL N < 2
END.
```

**Listing 4.10:** Solution of linear equations by the Gauss-Seidel method (continued).

## *Running the Gauss-Seidel Program*

Make a copy of Listing 4.2, and give it the name GAUSID.PAS. Change it to look like Listing 4.10. When you run this program, you will be asked for the number of equations. Enter the value 3. Then enter the three equations from the electrical circuit of Figure 4.1. The order of the equations is now immaterial since we have incorporated a row-interchange routine. You will next be asked for the relaxation factor. Give a value of 1. Each successive iteration will be displayed after the iteration number.

Convergence will occur after about 20 iterations. The program will again ask for the number of equations. Give a value of $-3$ this time. The minus sign indicates that the equations from the previous step are to be reused.

You can repeatedly run the program, trying different values for the relaxation factor. The following table shows the dependence of number of iterations on the choice of the relaxation factor:

| Lambda | Iterations |
|--------|------------|
| 0.8    | 31         |
| 1.0    | 20         |
| 1.2    | 11         |
| 1.3    | 9          |
| 1.4    | 12         |
| 1.5    | 15         |
| 1.8    | 44         |

Next enter the 2-by-2 Hilbert matrix

$$\begin{bmatrix} 1.0 & 0.5 \\ 0.5 & 0.3333 \end{bmatrix} \quad \begin{bmatrix} 1.5 \\ 0.83333 \end{bmatrix}$$

You will find that the best relaxation factor occurs at a value of 1.4, about the same as for the previous set of equations. Yet, for other sets of equations, the best value of Lambda might be 1.0 or 0.8.

Obviously, the Gauss-Seidel method is not as automatic a technique as the others that we have considered. However, you should consider using it to solve large numbers of linear equations or for solving sets of nonlinear equations.

## Summary

We have studied several methods for solving simultaneous equations, each method suited to a different situation. We have presented Turbo Pascal programs to carry out the algorithms of each of these methods. We have also investigated several special cases: multiple constant vectors, ill-conditioned equations, best-fit solutions for an overdetermined equation system, and equations with complex coefficients. We also wrote a version to read the data from a disk file. In the programs for these special examples, we have seen an abundance of new and powerful features of Turbo Pascal programming.

## Exercises

**4-1.** Show that the solution to the following set of linear equations:

$$4X_1 + 3X_2 + 2X_3 + X_4 = 10$$
$$-X_1 + 4X_2 - 2X_3 - X_4 = 0$$
$$-2X_2 + 4X_3 + 3X_4 = 5$$
$$-2X_1 - X_2 + X_3 + 4X_4 = 2$$

is $[1 \quad 1 \quad 1 \quad 1]$.

**4-2.** Show that the solution to the equations:

$$5X_1 + 2X_2 + X_3 + 3X_4 = 6$$
$$4X_1 + 3X_2 + X_3 - 2X_4 = -4$$
$$3X_1 + X_2 + 2X_3 + X_4 = 2$$
$$-X_1 + X_2 + 5X_3 + 3X_4 = -7$$

is $[2 \quad -3 \quad -1 \quad 1]$.

**4-3.** Interchange the inductor and the capacitor in Figure 4.8 and find the new loop currents.

# Development of a Curve-Fitting Program

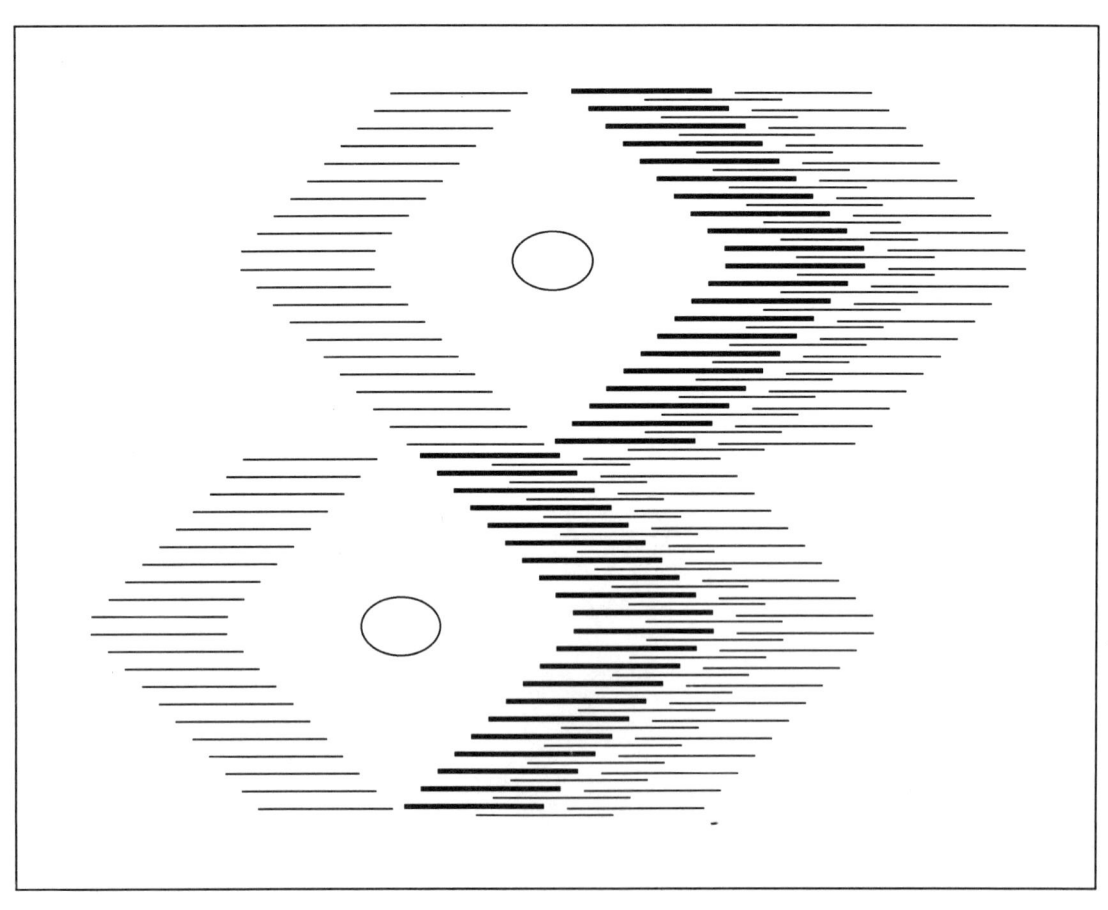

*Chapter* **5**

In this chapter, we will derive the expressions needed for least-squares curve-fitting. In particular, we will develop a computer program for finding the best straight line that can represent a set of x-y data. This program generates the data, calculates the desired equation, displays the results, plots the data, and supplies a measure of the correlation between the variables X and Y. We will add a sorting routine in Chapter 6 so that we can use this program with experimental data.

Instead of writing the entire program at one time, we will use a modular, top-down approach. This allows us to check each step along the way.

As we develop the different parts of this program, we will discuss the following algorithms and see how to program them in Turbo Pascal:

- Use of the Random function to simulate scattered-line experimental data

- A procedure for plotting graphs on a regular character-oriented video screen or on a printer

- A least-squares curve-fitting procedure, using differential calculus to arrive at the slope and y-intercept of the fitted data

- A simple and elegant method for integrating the correlation coefficient into our program

# *The Main Program*

First we will write the main program with the input and output routines. The main program will contain as little as possible: the program name, the declaration statements, and the calls to the various procedures.

## *First Version: Using the Built-in Random Function*

Create the Turbo Pascal source program shown in Listing 5.1. Use the file name CFIT1.PAS for the first version.

The main program begins with the program name and the declaration of the necessary global variables. Then two procedures are added. One procedure, Get_Data, reads data from the keyboard, and the other, Write_Data, displays information on the video screen. We place the calls to these procedures near the end of the main program. Notice that a semicolon follows the last procedure call. This is not necessary, but including it now will simplify our next programming step.

The input procedure Get_Data in this version generates a set of x-y points using Turbo Pascal's built-in random-number generator. Later, we will alter procedure Get_Data so that it will read the experimental data from a disk file rather than from the keyboard.

```
PROGRAM Cfit1;
{ Turbo Pascal program to perform a }
{ linear least-squares fit }
{USES Crt;}  { Use for versions 4 and later }

CONST
  Max = 20;

TYPE
  Ary = ARRAY[1..Max] OF Real;

VAR
  X, Y, Y_Calc: Ary;
  N: Integer;
  Done: Boolean;
  A, B: Real;

PROCEDURE Get_Data(VAR X, Y: Ary;
                   VAR N: Integer);
{ Get values for N and arrays X, Y }
{ Y is randomly scattered about a straight line }
```

**Listing 5.1:** The beginning of a curve-fitting program.

```
CONST
  A = 2.0;
  B = 5.0;

VAR
    I, J: Integer;
  Factor: Real;

BEGIN
  Write('Factor? ');
  ReadLn(Factor);
  IF Factor < 0.0 THEN
    Done := True
  ELSE
    BEGIN
      REPEAT
        Write('How many points? ');
        ReadLn(N)
      UNTIL (N > 2) AND (N <=  Max);
      FOR I := 1 TO N DO
        BEGIN
          J := N +1 -I;
          X[I] := J;
          Y[I] :=
            (A + B*J)*(1.0 +(2.0*Random-1.0)*Factor)
        END   { FOR loop }
    END   { IF }
END; { procedure Get_Data }

PROCEDURE Write_Data;
{ print out the answers }

VAR
  I: Integer;

BEGIN
  WriteLn;
  WriteLn('  I       X         Y');
  FOR I := 1 TO N DO
    WriteLn(I:3, X[I]:8:1, Y[I]:9:2);
  WriteLn
END; { Write_Data }

BEGIN   { main program }
  Done := False;
  REPEAT
    Get_Data(X, Y, N);
    IF NOT Done THEN
      BEGIN
        Write_Data;
        { more lines to be added here }
      END
  UNTIL Done
END.
```

**Listing 5.1:** The beginning of a curve-fitting program (continued).

Let us now look more closely at procedure Get_Data and its algorithm for simulating experimental data.

## *The Scattering Algorithm*

Procedure Get_Data generates a straight line with an intercept (A) of 2 and a slope (B) of 5. That is, it generates a set of data in the arrays X and Y, corresponding to the line $Y = 2 + 5X$.

We then use the random-number generator to move the points off the line according to the variable Factor. If the value of Factor is 0, the procedure generates a straight line. On the other hand, if Factor has a value of 0.2, the points will be displaced to a maximum of 20 percent from the line.

The random-number generator creates uniformly distributed random numbers. Although it would be more realistic to generate Gaussian random numbers because they more closely match experimental data, it is much faster to use the uniformly distributed ones. Furthermore, we will be removing this part of the program when our data are incorporated into the curve-fitting routine.

The scattering algorithm works in the following way:

1. Function Random returns a real number between 0 and 1.

2. The value is doubled to give a range of 0 to 2.

3. The subtraction of 1 sets the range from $-1$ to $+1$.

4. The result is multiplied by Factor and added to 1 to give the desired range.

Notice that X, Y, and N are passed to procedure Get_Data through a formal parameter list. Although this approach is not necessary in this case since the procedure could obtain the values globally, it does serve to isolate the input data from the main program. Suppose, for example, that after reading values for X and Y, we wanted to make a transformation. We could take the reciprocal of X and the logarithm of Y. Our equation would then be

$$\text{Ln } Y = A + \frac{B}{X}$$

The dummy parameter list of Get_Data could then contain the transformed values. The procedure heading might look like this:

```
Procedure Get_Data (VAR XR, LogY : Ary;VAR N : Integer);
```

By contrast, procedure Write_Data has no parameters.

## *Running the Main Program*

Run the main program to try it out. When the program asks you for a value for Factor, enter a 0 the first time. When it asks for the number of points, enter 9. The result will be three columns of numbers, as shown in Figure 5.1. Notice that the elements of array X are generated in descending order. In the next chapter, we will write several routines that can reorder the arrays X and Y.

Next the program asks for another value for Factor. This time, respond with 0.2. The resulting values for X should be the same as the previous calculations. The values for Y, however, will be somewhat larger or smaller because the random-number generator has moved them off the original straight line. When the program completes this step, it asks for another value for Factor. Enter a negative number or press Ctrl-C to terminate the program.

Now that we have a working program, we can begin to add new features. But before we make any additions, make a copy of the original version and then work with that copy. Continue to make copies of each revised version. Then if you have trouble with a new version, you can return to the previous one and start again.

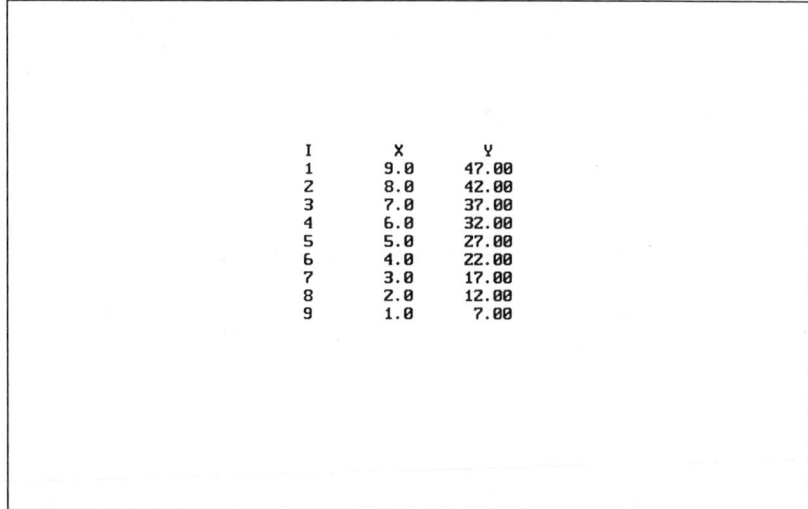

```
I       X       Y
1      9.0    47.00
2      8.0    42.00
3      7.0    37.00
4      6.0    32.00
5      5.0    27.00
6      4.0    22.00
7      3.0    17.00
8      2.0    12.00
9      1.0     7.00
```

**Figure 5.1:** First run of the curve-fitting program with Factor 0.

# *Programming a Plotter Routine*

Next we will add a routine for plotting the results on a character-oriented video screen. With a digital plotter and a video screen, it is possible to display data to a high degree of precision. The color/graphics screen (CGS), the extended graphics screen (EGS), and the Hercules graphics adapter can display high-quality graphics, but they require special routines to do so. (Borland International provides a routine for a high-quality graphics display with the Turbo Graphix Toolbox.) On the other hand, monochrome screens cannot display high-quality graphics. In this book, we will use an approach that is applicable to all screen types; that is, we will not use high-quality graphics. However, you can easily add this feature if you have a graphics display and the appropriate software.

As an alternative to high-quality graphics, we will display the experimental data on any video screen or printer by using the regular plus ( + ) and asterisk (*) symbols. The resulting plot will be a crude representation of the data, but it will suffice for revealing incorrect data and gross errors in programming.

A Turbo Pascal procedure for plotting one or two dependent variables as a function of a third independent variable is shown in Listing 5.2. Make a copy of this routine, and give it the file name PLOT.PAS. This routine plots the independent variable vertically rather than in the usual horizontal direction. The dependent variables are then displayed horizontally; that is, the graph is rotated clockwise one-quarter turn from the usual axis orientation. We will use this plot to display simple functions. However, the routine can also display multivalued functions. Furthermore, the values of the independent variable need not be uniformly spaced.

## *Adding the Plotter Routine to the Main Program*

Make a copy of the main program, and give it the name CFIT2.PAS. We will add two lines of code so that procedure Plot can display the data. First add the line

```
{$I PLOT.PAS}
```

just before the line

```
BEGIN { main program }
```

```
PROCEDURE Plot(        { with arrays }
                  X, { as independent variable }
                  Y, { as dependent variable }
                  Ycalc { as fitted curve } : Ary;
        { and }  M: Integer { number of points });

{ plot Y and Ycalc as a function of X for M points }
{ if M is negative, only X and Y are plotted }

CONST
  Blank = ' ';
  Line1 = 51;

VAR
  Ylabel: ARRAY[1..6] OF Real;
  Out:    ARRAY[1..Line1] OF Char;
  Lines, I, J, Jp, L, N: Integer;
  Iskip, Yonly: Boolean;
  Xlow, Xhigh, Xnext, Xlabel, Xscale, SignXs,
  Ymin, Ymax, Change, Yscale, Ys10: Real;

FUNCTION Pscale(P: Real): Integer;

BEGIN
  Pscale := Trunc((P - Ymin) / Yscale + 1)
END; { Pscale }

PROCEDURE Outlin(Xname: Real);
{ output a line }

VAR
  I, Max: Integer;

BEGIN
  Write(Xname:8:2, Blank); { line label }
  Max := Line1 + 1;
  REPEAT          { skip blanks on end of line }
    Max := Max - 1
  UNTIL (Out[Max] <> Blank) OR (Max = 1);
  FOR I := 1 TO Max DO
    Write(Out[I]);
  WriteLn;
  FOR I := 1 TO Max DO
    Out[I] := Blank        { Blank next line }
END; { Outlin }

PROCEDURE Setup(Index: Integer);
{ set up the plus and asterisk for printing }

CONST
  Asterisk = '*';
  Plus = '+';

VAR
  I: Integer;
```

**Listing 5.2:** A plotting procedure.

```
BEGIN
  I := Pscale(Y[Index]);
  Out[I] := Plus;
  IF NOT Yonly THEN
    BEGIN           { add Ycalc too }
      I := Pscale(Ycalc[Index]);
      Out[I] := Asterisk
    END
END; { Setup }

BEGIN              { body of plot }
  IF M > 0 THEN { Plot Y and Ycalc vs X }
    BEGIN
      N := M;
      Yonly := False
    END
  ELSE             { Plot only Y vs X }
    BEGIN
      N := - M;
      Yonly := True
    END;
  { space out alternate lines }
  Lines := 2 * (N - 1) + 1;
  WriteLn;
  Xlow := X[1];
  Xhigh := X[N];
  Ymax := Y[1];
  Ymin := Ymax;
  Xscale := (Xhigh - Xlow) / (Lines - 1);
  SignXs := 1.0;
  IF Xscale < 0.0 THEN SignXs := -1.0;
    FOR I := 1 TO N DO
    BEGIN
      IF Y[I] < Ymin THEN  Ymin := Y[I];
      IF Y[I] > Ymax THEN  Ymax := Y[I];
      IF NOT Yonly THEN
        BEGIN
          IF Ycalc[I] < Ymin THEN
            Ymin := Ycalc[I];
          IF Ycalc[I] > Ymax THEN
            Ymax := Ycalc[I]
        END       { if Yonly }
    END;
  Yscale := (Ymax - Ymin) / (Linel - 1);
  Ys10 := Yscale * 10;
  Ylabel[1] := Ymin;        { Y axis }

FOR I := 1 TO 4 DO
  Ylabel[I + 1] := Ylabel[I] + Ys10;
Ylabel[6] := Ymax;
FOR I := 1 TO Linel DO
  Out[I] := Blank;        { blank line }
Setup(1);
L := 1;
Xlabel := Xlow;
Iskip := False;
```

Listing 5.2: A plotting procedure (continued).

```
    FOR I := 2 TO Lines DO  { set up a line }
      BEGIN
        Xnext := Xlow + Xscale * (I - 1);
        IF Iskip THEN
          WriteLn('   -')
        ELSE
          BEGIN
            L := L + 1;
            WHILE
              (X[L] - (Xnext - 0.5 * Xscale))*SignXs
                <= 0.0    DO
              BEGIN
                Setup(L);  { set up print line }
                L := L + 1
              END; { WHILE }
            Outlin(Xlabel);        { print a line }
            FOR J := 1 TO Line1 DO
              Out[J] := Blank           { blank line }
          END;        { if Iskip }
        IF (X[L] - (Xnext + 0.5 * Xscale))*SignXs > 0.0
          THEN Iskip := True
        ELSE
          BEGIN
            Iskip := False;
            Xlabel := Xnext;
            Setup(L)       { Setup print line }
          END
    END;            { FOR loop }
  Outlin(Xhigh); { last line }
  Write('   ');
  FOR I := 1 TO 6 DO
    Write('       ^    ');
  WriteLn;
  Write('   ');
  FOR I := 1 TO 6 DO
    Write(Ylabel[I]:9:1, Blank);
  WriteLn;
  WriteLn
END; { Plot }
```

**Listing 5.2:** A plotting procedure (continued).

Then, just after the line

```
Write_Data;
```

place the lines

```
WriteLn(' Press any key for plot');
REPEAT
UNTIL KeyPressed;
Plot (X, Y, Y, -N);
```

The second and third lines freeze the screen until you can read the first part of the output. Then, when you press any keyboard key, the output from the plot routine is displayed. KeyPressed is a logical function built into Turbo Pascal to sample the keyboard. If a key is pressed, it returns the value True; otherwise it returns the value False. The fourth line calls the plotting routine.

We are going to use the procedure Plot in all the remaining routines in this chapter, as well as those in later chapters. By referring to this procedure with the INCLUDE directive, we do not have to add a separate copy of the procedure in each program.

The lower portion of your program should now look like Listing 5.3. As before, the semicolon at the end of this version is not necessary now, but it will simplify our next addition to the program. Notice that the second and third arguments to Plot are the same and that the last argument is negative. When the last argument of Plot is negative, it means that only a single curve is to be plotted.

## *Running the New Version*

Run this version to try it out. Enter a value of 0.2 for Factor, and the plotter output will look like Figure 5.2. If you enter a value of 0 for Factor, you will get a nearly straight line.

```
{$I PLOT.PAS} {Listing 5.2}

BEGIN   { main program }
  Done := False;
  REPEAT
    Get_Data (X, Y, N);
    IF NOT Done THEN
      BEGIN
        Write_Data;
        WriteLn(' Press any key for plot');
        REPEAT
        UNTIL KeyPressed;
        Plot(X, Y, Y, -N);
        { more lines to be added here }
      END
  UNTIL Done
END.
```

**Listing 5.3:** Adding a call to procedure Plot.

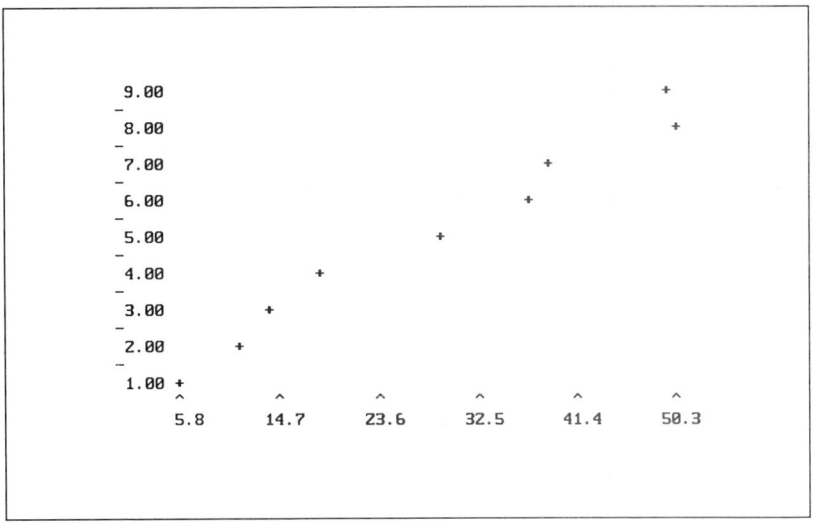

**Figure 5.2:** Y plotted as a function of X.

To test the capabilities of our plotter routine, we will have it plot two curves at once. To do this, we need to incorporate the small procedure Linfit into our program to simulate a linear fit. This version of Linfit will be replaced by a much more sophisticated procedure later in the chapter, after we have studied the least-squares curve-fitting algorithm.

## *A Simulated Curve Fit*

For our next step, then, we will add a simulated curve-fitting routine. This procedure will simply generate a vector Y_Calc that lies on our straight line with a slope of 5 and an intercept of 2. Keep in mind that we will add a proper curve-fitting routine in a later version.

Make a copy of your current program giving it the name CFIT3.PAS. Add procedure Linfit, shown in Listing 5.4, to your source program. Place it immediately after procedure Write_Data. We must also make some changes to the main program and to procedure Write_Data. The current version of the main program is shown in Listing 5.6, and the new procedure Write_Data is shown in Listing 5.5. Notice that Y_Calc is now the third parameter in the call to Plot and that the final parameter is positive.

## *Running the Plotter Routine with Two Curves*

Now we will use the new version to plot two curves. As before, the program will ask you to enter a value for Factor. Respond with an answer of 0.2. This will produce four columns of data, including the values for the calculated data, Y_Calc. The graph that follows the tabular data shows two different sets of data. Asterisks are used to

```
PROCEDURE Linfit(X, Y: Ary;
            VAR Y_Calc: Ary;
            VAR A, B  : Real;
                N: Integer);
{ generate a straight line for X-Y }

VAR
  I: Integer;

BEGIN  { Linfit }
   A := 2.0;
   B := 5.0;
   FOR I := 1 TO N DO
      Y_Calc[I] := A + B * X[I]
END;  { Linfit }
```

**Listing 5.4:** Procedure Linfit to simulate a linear fit.

```
PROCEDURE Write_Data;
{ print out the answers }

VAR
  I: Integer;

BEGIN
  WriteLn;
  WriteLn(' I       X        Y      Y CALC');
  FOR I := 1 TO N DO
    WriteLn(I:3, X[I]:8:1, Y[I]:9:2, Y_Calc[I]:9:2);
  WriteLn
END; { Write_Data }
```

**Listing 5.5:** The revised procedure Write_Data.

represent the values of Y_Calc. They will form a nearly perfect straight line. In addition, plus signs represent the values of Y. They will be scattered on both sides of the Y_Calc values, as shown in Figure 5.3. If the two symbols are coincident for a given value of X, only the asterisk is shown.

```
BEGIN  { main program }
  Done := False;
  REPEAT
    Get_Data(X, Y, N);
    IF NOT Done THEN
      BEGIN
        Linfit(X, Y, Y_Calc, A, B, N);
        Write_Data;
        WriteLn(' Press any key for plot');
        REPEAT
        UNTIL KeyPressed;
        Plot(X, Y, Y_Calc, N)
      END
  UNTIL Done
END.
```

**Listing 5.6:** The current main program.

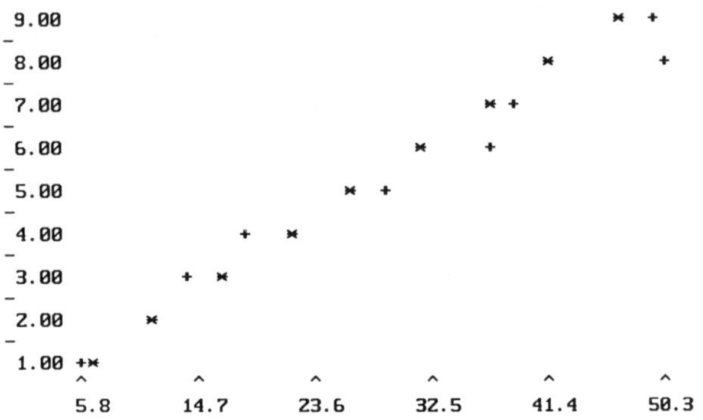

**Figure 5.3:** Y and Y_Calc plotted against X.

When the program asks for another value of Factor, respond with a 0. Now both Y and Y_Calc will have the same values. Since the two lines are coincident, only one line is shown. Finally, give a negative value for Factor, or press Ctrl-C to terminate the program.

Now that we have written and tested both the main program and a procedure for plotting curves, we are ready to move on to the main topic of this chapter.

## *The Curve-Fitting Algorithm*

So far, we have simply generated the values of Y_Calc that correspond to our original line. The time has come to derive the algorithm for a linear, least-squares procedure. We will first introduce a new vector, R, which contains the *residuals*.

For each experimental point, corresponding to an x-y pair, there will be an element of R that represents the difference or residual between the calculated value Y_Calc (which we will represent as $\hat{Y}$) and the corresponding original value of Y. This can be expressed mathematically as

$$R_i = \hat{Y}_i - Y_i \tag{1}$$

Although a point will occasionally coincide with the calculated curve, in general, about half of the x-y points will lie on one side of the fitted curve, resulting in positive values for R, and the remaining points will lie on the other side of the curve, giving negative values for R. The sum of these residuals will be close to zero.

The least-squares, curve-fitting criterion is that the sum of the squares of the residuals be minimized. The square of each residual is positive, therefore the sum of the residuals squared (known as SRS) will be a positive number. This criterion can be expressed as

$$SRS = \sum_{i=1}^{N} R_i^2 = \text{minimum} \tag{2}$$

where N is the number of x-y points (and the length of the vectors X, Y, and $\hat{Y}$).

By combining Equation 1 with the curve-fitting equation

$$\hat{Y}_i = A + BX_i \tag{3}$$

we get

$$R_i = A + BX_i - Y_i \tag{4}$$

and

$$SRS = \sum_{i=1}^{N} R_i^2 = \sum_{i=1}^{N} (A + BX_i - Y_i)^2 \tag{5}$$

The problem is reduced to finding the values of A and B so that the summation of Equation 5 is a minimum. We accomplish this with differential calculus. We take the derivative of Equation 5 with respect to each variable (A and B in this case) and set the result to zero.

$$\frac{\partial \Sigma R_i^2}{\partial A} = 0 \quad \text{and} \quad \frac{\partial \Sigma R_i^2}{\partial B} = 0 \tag{6}$$

Substituting Equation 5 into Equation 6 gives

$$\frac{\partial \Sigma (A + BX_i - Y_i)^2}{\partial A} = 0 \tag{7}$$

and

$$\frac{\partial \Sigma (A + BX_i - Y_i)^2}{\partial B} = 0 \tag{8}$$

which is equivalent to

$$\frac{2\Sigma (A + BX_i - Y_i)\partial \Sigma (A + BX_i - Y_i)}{\partial A} = 0 \tag{9}$$

and

$$\frac{2\Sigma (A + BX_i - Y_i)\partial \Sigma (A + BX_i - Y_i)}{\partial B} = 0 \tag{10}$$

Since B, X, and Y are not functions of A, and the derivative of A with respect to itself is unity, Equation 9 reduces to

$$\sum A + \sum BX_i = \sum Y_i \tag{11}$$

Similarly, A, X, and Y, are not functions of B. Therefore, Equation 10 becomes

$$\sum AX_j + \sum BX_i^2 = \sum X_i Y_i \tag{12}$$

A and B are constants, and so they can be factored from the summation step. Equations 7 and 8 can then be expressed as

$$AN + B\sum X_i = \sum Y_i \tag{13}$$

and

$$A\sum X_i + B\sum X_i^2 = \sum X_i Y_i \tag{14}$$

We have thus reduced the problem of finding a straight line through a set of x-y data points to one of solving two simultaneous equations (13 and 14). Both these equations are linear in the unknowns A and B. (X, Y, and N are the original data.) The simultaneous solution can be obtained by using Cramer's rule, as follows:

$$A = \frac{\begin{vmatrix} \sum Y_i & \sum X_i \\ \sum X_i Y_i & \sum X_i^2 \end{vmatrix}}{\begin{vmatrix} N & \sum X_i \\ \sum X_i & \sum X_i^2 \end{vmatrix}} \tag{15}$$

and

$$B = \frac{\begin{vmatrix} N & \sum Y_i \\ \sum X_i & \sum X_i Y_i \end{vmatrix}}{\begin{vmatrix} N & \sum X_i \\ \sum X_i & \sum X_i^2 \end{vmatrix}} \tag{16}$$

The corresponding equations we have to solve are

$$A = \frac{\sum X_i^2 \sum Y_i - \sum X_i \sum X_i Y_i}{N\sum X_i^2 - \sum X_i \sum X_i} \tag{17}$$

and

$$B = \frac{N\sum X_i Y_i - \sum X_i \sum Y_i}{N\sum X_i^2 - \sum X_i \sum X_i} \tag{18}$$

The computer calculation of A and B is straightforward. The summation of X is obtained by summing the values of X. The summation of $X^2$ is obtained by squaring each value of X and then adding up the squares.

Equations 17 and 18 are commonly converted into an equivalent form by dividing the numerator and denominator by N

$$A = \frac{(\sum X_i^2 \sum Y_i - \sum X_i \sum X_i Y_i)/N}{\sum X_i^2 - \sum X_i \sum X_i/N} \tag{19}$$

and

$$B = \frac{\sum X_i Y_i - \sum X_i \sum Y_i/N}{\sum X_i^2 - \sum X_i \sum X_i/N} \tag{20}$$

The denominators of Equations 19 and 20 appear in the formula for standard deviation presented in Chapter 2.

We have outlined the mathematics of least-squares curve-fitting and have derived formulas for finding the slope (B) and y-intercept (A) of a linear-fitted curve. Now we are ready for the real Linfit procedure.

## *The Curve-Fitting Procedure*

A procedure for fitting a straight line is shown in Listing 5.7. Make a copy of your previous program, and give it the name CFIT4. Delete the original procedure Linfit, and replace it with the new one shown in Listing 5.7.

```
PROCEDURE Linfit(X, Y: Ary;
             VAR Y_Calc: Ary;
             VAR A, B  : Real;
                    N: Integer);

{ fit a straight line (Y_Calc) through
   N sets of X and Y pairs of points }

VAR
   I: Integer;
   Sum_X, Sum_Y, Sum_XY, Sum_X2,
   Sum_Y2, Xi, Yi, SXY, SXX, SYY: Real;

BEGIN  { Linfit }
   Sum_X  := 0.0;
   Sum_Y  := 0.0;
   Sum_XY := 0.0;
   Sum_X2 := 0.0;
   Sum_Y2 := 0.0;
   FOR I := 1 TO N DO
      BEGIN
         Xi := X[I];
         Yi := Y[I];
         Sum_X  := Sum_X + Xi;
         Sum_Y  := Sum_Y + Yi;
         Sum_XY := Sum_XY + Xi * Yi;
         Sum_X2 := Sum_X2 + Xi * Xi;
         Sum_Y2 := Sum_Y2 + Yi * Yi
      END;
   SXX := Sum_X2 - Sum_X * Sum_X/ N;
   SXY := Sum_XY - Sum_X * Sum_Y/ N;
   SYY := Sum_Y2 - Sum_Y * Sum_Y/ N;
   B := SXY / SXX;
   A :=((Sum_X2 * Sum_Y - Sum_X * Sum_XY)/ N) / SXX;
   FOR I := 1 TO N DO
      Y_Calc[I] := A + B * X[I]
END; { Linfit }
```

**Listing 5.7:** Procedure Linfit to generate a least-squares fit.

At the beginning of procedure Linfit, the variables Sum_X, Sum_Y, etc., are set to 0, and the FOR loop is used to calculate the desired sums. Notice that a change of variable is made at the beginning of the loop:

```
Xi := X[I];
Yi := Y[I];
```

This change shortens formulas that contain these variables; however, it does not seem to change the speed of Turbo Pascal.

The arrays X, Y, and Y_Calc, and the scalar variables A, B, and N, are formal parameters as before. With this arrangement, we could call Linfit to fit X and Y at one point in the program:

```
Linfit (X, Y, Y_Calc, A, B, N);
```

Then, at a later point, we could again call Linfit to fit X1 and Y1 with a different number of points:

```
Linfit (X1, Y1, Y1c, A1, B1, N);
```

For the second call to Linfit, a straight line with slope B1 and intercept A1 is fitted through the X1, Y1 pairs.

## Running the Curve-Fitting Program

Run the new version to try it out. Give Factor a value of 0.2. The fitted line of asterisks should go neatly through the scattered plus signs. Compare Figure 5.4, which shows the actual fit, with Figure 5.3, which gives the simulated fit. Next, give Factor a value of 0. Only a single line of asterisks is apparent now. Furthermore, the intercept should be equal to 2 and the slope should be equal to 5, the initial values.

# The Correlation Coefficient

Although we now have a procedure for calculating the straight line, we are not finished yet. We can obtain the equation of a line fitting our experimental data and then use it to predict a value for Y from a given value of X. Under certain circumstances, however, our mathematically correct solution is useless.

Consider, for example, the set of data shown in Figure 5.5. Our curve-fitting procedure Linfit can find the equation of a straight line through the data, but the resulting line does not give us any additional information. That is, a knowledge of the behavior of X does not tell us anything about the behavior of Y. There is no correlation between X and Y.

Figure 5.6 shows another case where a knowledge of X is no help in predicting the behavior of Y. Again, there is no correlation between X and Y.

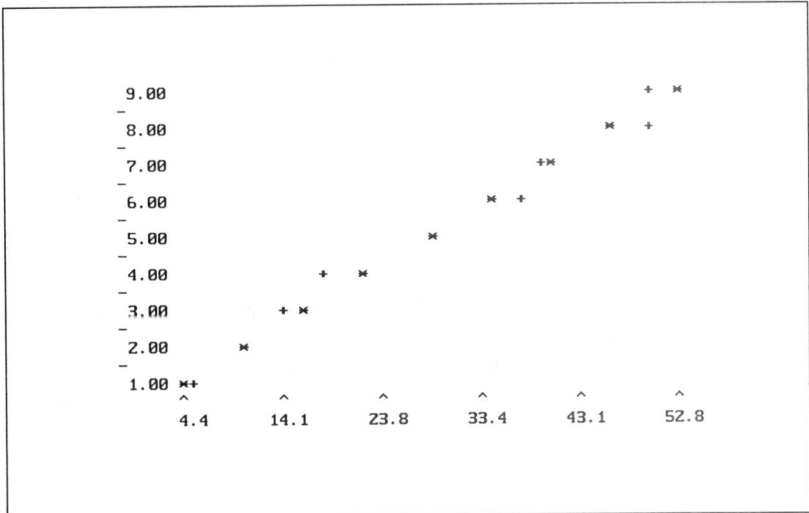

**Figure 5.4:** A least-squares fit to Y versus X.

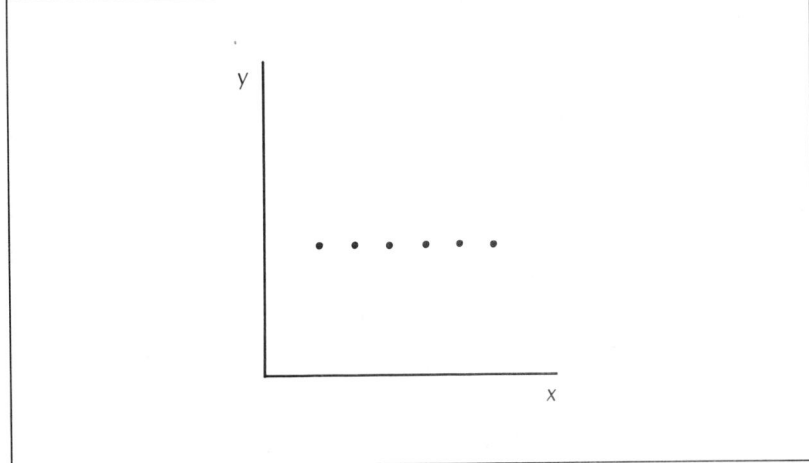

**Figure 5.5:** No correlation between X and Y.

We need to obtain a quantitative measure of the correlation between X and Y. We want to know how well we can predict the behavior of Y if we know the behavior of X. The measure we need is the *correlation coefficient*.

We saw in Chapter 2 that we could characterize a set of data by the mean and the standard deviation of the values about their own mean. We can also calculate the standard deviation of Y about the fitted curve. This measure is termed the *standard error of the estimate* (SEE). The correlation coefficient compares the standard deviation of Y (about its own mean) to the standard deviation about the fitted curve.

The correlation coefficient is 0 when there is no correlation, as in the examples shown in Figures 5.5 and 5.6. On the other hand, the correlation coefficient approaches unity as the data approach a straight line. The correlation coefficient for the data given in Figure 5.4 is 0.99.

We will now incorporate calculations for the correlation coefficient into our program.

## Programming Correlation Coefficient Calculations

With a few additional statements, our program can calculate the correlation coefficient and the standard errors in the coefficients (the intercept and slope). The standard errors are standard deviations for the

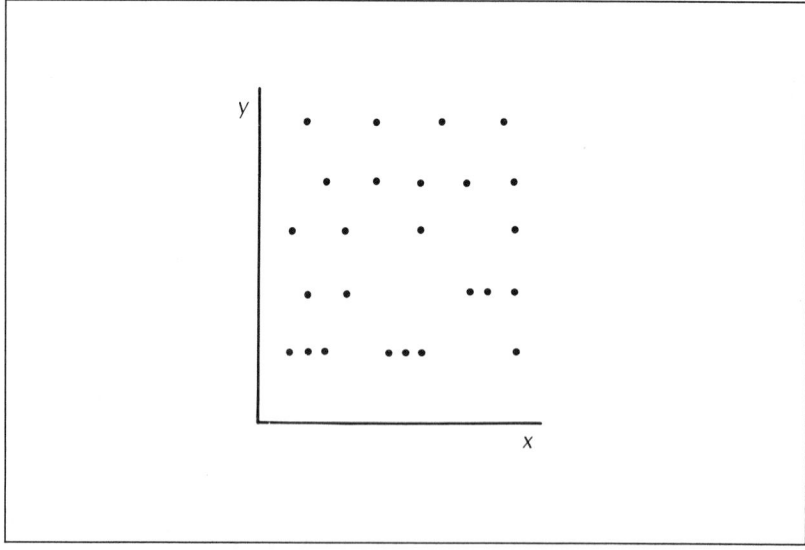

**Figure 5.6:** Badly scattered data.

coefficients. They can be used to determine confidence intervals for each coefficient.

Make a copy of the previous version, and give it the name CFIT5.PAS. For the next changes, refer to the complete listing given in Listing 5.8. First add the expression

Correl_Coef, Sigma_A, Sigma_B, SEE : Real;

to the VAR section at the beginning of the main program. These will then be global variables. Place the next four lines in procedure Linfit, immediately after the definition of the intercept A.

```
Correl_Coef :=
    SXY/ Sqrt(SXX * SYY);
SEE :=
    Sqrt( (Sum_Y2 − A*Sum_Y − B*Sum_XY) / (N−2));
Sigma_B := SEE / Sqrt(SXX);
Sigma_A := Sigma_B * Sqrt( Sum_X2 / N);
```

```
PROGRAM Cfit5;
{ Turbo Pascal program to perform a }
{ linear least-squares fit }
{USES Crt;}   { Use for versions 4 and later }

CONST
  Max = 20;

TYPE
  Ary = ARRAY[1..Max] OF Real;

VAR
  X, Y, Y_Calc: Ary;
  N: Integer;
  Done: Boolean;
  A, B, Correl_Coef,
  Sigma_A, Sigma_B, SEE: Real;

PROCEDURE Get_Data(VAR X, Y: Ary;
                   VAR N: Integer);
{ Get values for N and arrays X, Y }
{ Y is randomly scattered about a straight line }

CONST
  A = 2.0;
  B = 5.0;
```

**Listing 5.8:** The complete curve-fitting program.

```
VAR
    I, J: Integer;
  Factor: Real;

BEGIN
  Write('Factor? ');
  ReadLn(Factor);
  IF Factor < 0.0 THEN
    Done := True
  ELSE
    BEGIN
      REPEAT
        Write('How many points? ');
        ReadLn(N)
      UNTIL (N > 2) AND (N <=  Max);
      FOR I := 1 TO N DO
        BEGIN
          J  := N +1 -I;
          X[I]  := J;
          Y[I] :=
            (A + B*J)*(1.0 +(2.0*Random-1.0)*Factor)
        END    { FOR loop }
    END   { ELSE }
END; { procedure Get_Data }

PROCEDURE Write_Data;
{ print out the answers }


VAR
  I: Integer;

BEGIN
  WriteLn;
  WriteLn(' I       X         Y      Y Calc');
  FOR I := 1 TO N DO
    WriteLn(I:3, X[I]:8:1,
            Y[I]:9:2, Y_Calc[I]:9:2);
  WriteLn;
  WriteLn(' Intercept is ', A:8:3,
          ', Sigma is ', Sigma_A:8:3);
  WriteLn('     Slope is ', B:8:2,
          ', Sigma is ', Sigma_B:8:3);
  WriteLn;
  WriteLn(' Correlation coefficient is ',
          Correl_Coef:7:4)
END; { Write_Data }

PROCEDURE Linfit(X, Y: Ary;
            VAR Y_Calc: Ary;
            VAR A, B  : Real;
                N: Integer);
{ fit a straight line (Y_Calc) through
  N sets of X and Y pairs of points }
```

**Listing 5.8:** The complete curve-fitting program (continued).

```
VAR
  I: Integer;
  Sum_X, Sum_Y, Sum_XY, Sum_X2,
  Sum_Y2, Xi, Yi, SXY ,SXX, SYY: Real;

BEGIN   { Linfit }
  Sum_X := 0.0;
  Sum_Y := 0.0;
  Sum_XY := 0.0;
  Sum_X2 := 0.0;
  Sum_Y2 := 0.0;
  FOR I := 1 TO N DO
    BEGIN
      Xi := X[I]; Yi := Y[I];
      Sum_X := Sum_X + Xi;
      Sum_Y := Sum_Y + Yi;
      Sum_XY := Sum_XY + Xi * Yi;
      Sum_X2 := Sum_X2 + Xi * Xi;
      Sum_Y2 := Sum_Y2 + Yi * Yi
    END;

  SXX := Sum_X2 - Sum_X * Sum_X/ N;
  SXY := Sum_XY - Sum_X * Sum_Y/ N;
  SYY := Sum_Y2 - Sum_Y * Sum_Y/ N;
  B := SXY / SXX;
  A :=
    ((Sum_X2 * Sum_Y - Sum_X * Sum_XY)/ N) / SXX;
  Correl_Coef :=
      SXY/ Sqrt(SXX * SYY);
  SEE :=
      Sqrt((Sum_Y2 - A*Sum_Y - B*Sum_XY) / (N-2));
  Sigma_B := SEE / Sqrt(SXX);
  Sigma_A := Sigma_B * Sqrt(Sum_X2 / N);
  FOR I := 1 TO N DO
    Y_Calc[I] := A + B * X[I]
END;   { Linfit }

{$I PLOT.PAS } {Listing 5.2}

BEGIN   { main program }
  Done := False;
  REPEAT
    Get_Data(X, Y, N);
    IF NOT Done THEN
      BEGIN
        Linfit(X, Y, Y_Calc, A, B, N);
        Write_Data;
        WriteLn(' Press any key for plot');
        REPEAT
        UNTIL KeyPressed;
        Plot(X, Y, Y_Calc, N)
      END
  UNTIL Done
END.
```

**Listing 5.8:** The complete curve-fitting program (continued).

Finally, change the output statements in procedure Write_Data to the following:

```
WriteLn(' Intercept is ', A:8:3,
          ', Sigma is ', Sigma_A:8:3);
WriteLn(' Slope is ', B:8:2,
          ', Sigma is ', Sigma_B:8:3);
WriteLn;
WriteLn(' Correlation coefficient is ',
        Correl_Coef:7:4)
```

The correlation coefficient adds a final test to our program—a way of quantifying the usefulness of the fitted curve. Now let us look at the finished program.

## Running the Finished Program

Run the current version of the curve-fitting program. First give Factor a value of 0. The intercept will be 2, and the slope will be 5, as before. In addition, the values of sigma for A and B will be 0, and the correlation coefficient will be equal to 1. The plot will show a straight line of asterisks.

Now try a Factor of 0.2. This will give sigma values greater than 0 for the intercept and the slope. The correlation coefficient will be somewhat less than unity.

## Summary

The modular development process that we have used for this program has allowed us to evaluate each of the procedures as we wrote them. We began with the main program and added the procedures step by step to simulate data, plot results, compute the fitted curve, and supply the correlation coefficient.

Although we now have a linear curve-fitting program that works, it is not very useful. It can only fit data produced by a random-number generator. Therefore, we will want to alter procedure Get_Data so that it can fit data that are embedded in the program or are input from the keyboard or from a disk file. We will delay this step, however, until after the next chapter, where we will add a sorting routine.

# *Exercises*

**5-1.** The vapor pressure of water is

| Temp (°C) | Pressure (torr) |
|-----------|-----------------|
| 20 | 17.535 |
| 30 | 31.824 |
| 40 | 55.324 |
| 50 | 92.51 |
| 60 | 149.38 |
| 70 | 233.7 |
| 80 | 355.1 |
| 90 | 525.76 |
| 100 | 760.00 |

where the temperature is given in degrees Celsius and the pressure is in torr (mm of mercury). Find the coefficients A and B to the equation

$$\text{Ln } P = A + B/T$$

by performing a least-squares fit on the above data. Be sure to convert temperature to kelvins by adding 273.15 to the Celsius value.

Make a copy of Listing 5.8 and alter the input procedure Get_Data. Remove the REPEAT/UNTIL loop and define the number of points:

```
N := 9;
```

The vector Y can be defined with direct statements:

```
Y[1] := 17.535
Y[2] := 31.824
. . .
```

as shown in Listing 7.1. Change the statements in the FOR loop so that the vector X is defined as the reciprocal of the kelvins temperature. Also take the logarithm of the Y values:

```
X(I) = 10.0 / ((I − 1) + 20 + 273.15)
Y(I) = Ln(Y(I))
```

**5-2.** Repeat Problem 5-1, but read the data from a disk file.

**5-3.** Incorporate a range check in the plot routine given in Listing 5.3 so that values of X and Y will be given in scientific E notation if they are very small or very large.

**5-4.** The plotter routine can only print one symbol at a particular location. Thus if a plus sign is coincident with an asterisk or another plus sign, only one symbol is displayed. Alter the plot routine so that an M symbol is displayed whenever there is more than one element at a particular location.

# Sorting

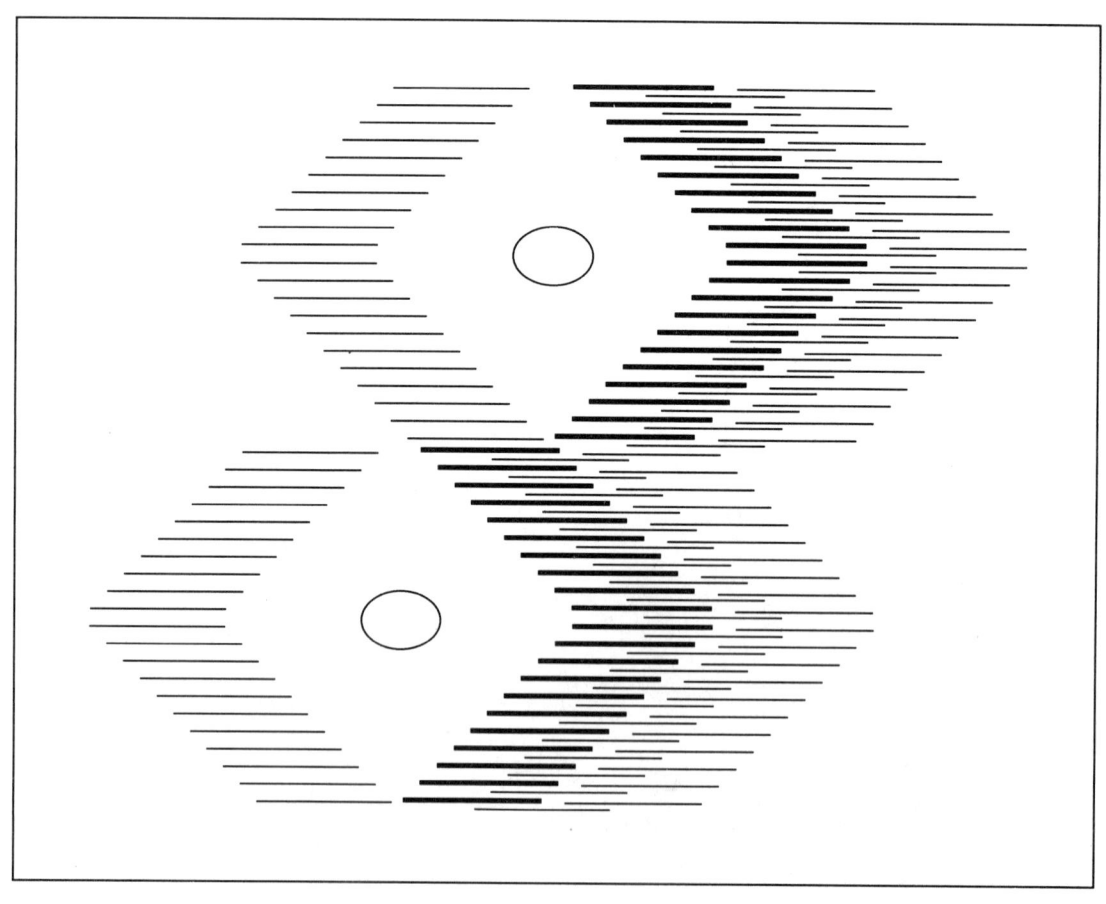

In this chapter, we will develop three different sorting algorithms: a bubble sort, a Shell sort, and a recursive quick sort. Then we will incorporate one of these routines into the curve-fitting program that we developed in Chapter 5.

First let us discuss the rationale for including a sort routine in our program.

## *Handling Experimental Data*

The curve-fitting program that we wrote in the previous chapter obtained its data from a random-number generator. In practice, your programs will get the data from the keyboard or from a disk file. Alternatively, the data can be embedded in the program itself.

In some cases, we might want to fit experimental data that have not been acquired in numeric order. For example, suppose that we want to investigate the thermal expansion of a material. We could take paired measurements of temperatures and the corresponding total sample lengths. To do this, we bring the experimental apparatus to a certain temperature and measure the length. Then we increase the temperature and measure the sample temperature and length again. However, it would not be wise to continue the experiment in this fashion because we might measure the length before the sample temperature became uniform, or before the sample reached the value of the temperature sensor. The resulting measured pairs would all contain errors in the same direction.

A better experimental technique would be to approach the desired temperature sometimes from below and sometimes from above. Then, the experimental data might look like this:

| Temperature | Length |
|---|---|
| 100 | 8.0 |
| 300 | 19.0 |
| 200 | 13.5 |
| 500 | 30.0 |
| 400 | 24.5 |

The curve-fitting program that we developed in the last chapter can readily find a least-squares fit to these data. We could place the data directly in procedure Get_Data in the following manner:

```
Procedure Get_Data . . .

. . .
BEGIN
N := 5
X[1] := 100; Y[1] := 8.0;
X[2] := 300; Y[2] := 19.0;
X[3] := 200; Y[3] := 13.5;
X[4] := 500; Y[4] := 30.0;
X[5] := 400; Y[5] := 24.5
END;
```

If we were to run the program with this addition, however, we would find that procedure Plot will not display the data correctly. The problem is that the array of independent variables, X, must be arranged either in increasing or decreasing order. That is, the data must be sorted. Procedure Plot worked properly in the previous chapter because the X array was generated in decreasing order. Notice that we are not concerned about the ordering of the dependent variable Y.

## Testing the Sorting Routines

There are many different sorting algorithms. Some are very fast; others are very slow. Some are faster with nearly sorted data, and some are slower under these conditions. Some require additional working space; others need only the space occupied by the original data.

We will write three sorting programs in this chapter, placing each in a different disk file. Then, by calling each one from a driver program, we will compare their performance.

The program shown in Listing 6.1 generates the data, calls the separate sorting routine, and then displays the results. When you run the program, it asks you for the number of items to be sorted. The Turbo Pascal random-number generator is then called to produce the desired number of elements. The program displays the original set of numbers on the video screen, ten numbers per line. The symbol Beep is defined as ASCII 7, which sounds the computer's beeper. We use it at the beginning and end of each sorting step so that we can compare the speed of the sorting routines.

At the end of the sorting step, the program sounds the beeper a second time and displays the sorted array on the screen. It also displays the word RANDOM because it uses a set of random numbers for this first phase. Check that the numbers are in order to ensure that the program is working properly.

```
PROGRAM Testsort;
{ test speed of sorting routine }

CONST
  Beep = #7;
  Max = 1000;

TYPE
  Sort_Var = Real;
  Ary = ARRAY[1..Max] OF Sort_Var;

VAR
      X: Ary;
  N, I: Integer;

PROCEDURE Print;

VAR
  I: Integer;

BEGIN
  WriteLn;
  FOR I := 1 TO N DO
    BEGIN
      Write(X[I]:7:0);
      IF (I MOD 10) = 0 THEN WriteLn
    END
END;

{$I SORT.PAS} {Listing 6.3, 6.4, or 6.5}
```

**Listing 6.1:** Program to test the sorting routines.

```
BEGIN   { main program }
   REPEAT
     REPEAT
       WriteLn;
       Write(' How many numbers? ');
       ReadLn(N)
     UNTIL N <= Max;
     FOR I :=1 TO N DO
       X[I] := 1000*Random;
     Print;
     Write(Beep);
     Sort(X, N); { random numbers }
     Write(Beep);
     Print;
     WriteLn(' Random');
     Write(Beep);
     Sort(X, N);  { sorted numbers }
     Write(Beep);
     Print;
     WriteLn(' sorted');
     FOR I := 1 TO N DO
       X[I] := N + 1 - I;
     Write(Beep);
     Sort(X, N);  { reversed numbers }
     Write(Beep);
     Print;
     WriteLn(' reversed')
   UNTIL N < 5
END.
```

**Listing 6.1:** Program to test the sorting routines (continued).

The beeper sounds a third time at the beginning of the second part of the test. This time, however, the procedure runs on the array that is already sorted. At the end of this second sorting step, the beeper sounds a fourth time, and the sorted array is shown again, along with the word SORTED.

For the third phase of the test program, an array of numbers is generated in reverse order. The program then calls the sorting routine for the third time. At the end of this step, it displays the sorted array and the word REVERSED.

## The Test and Swap Routines

Type the main program shown in Listing 6.1, and give it the name TESTSORT.PAS. Notice that this program uses the INCLUDE directive to refer to a separate file named SORT.PAS.

Create a second file called SWAP.PAS that holds the routine given in Listing 6.2. This routine interchanges (swaps) two items that are out of sequence. We will place a copy of this file at the beginning of each sorting routine.

```
PROCEDURE Swap(VAR P, Q: Sort_Var);

VAR
  Hold: Sort_Var;

BEGIN
  Hold := P;
  P  := Q;
  Q  := Hold
END;          { Swap }
```

**Listing 6.2:** Procedure Swap to interchange values.

# *The Bubble Sort*

The first sorting algorithm that we will consider is known as the *bubble sort*. This routine is the easiest to understand and the easiest to program, but, unfortunately, it is also the slowest. For sorting lists that contain fewer than two dozen items, however, speed is not really important.

With the bubble sort, each element is compared with all the remaining, unsorted elements. If a particular pair is found to be out of order, the two elements are interchanged. There are two loops, one nested inside the other. The outer loop runs from 1 to one less than the length of the array. The inner loop runs from one larger than the outer loop up to the length of the array. With this algorithm, the smaller items "float" to the top of the array during the sorting process (this is the origin of the name bubble sort).

## *Programming the Bubble Sort*

Create a third file named SORTB.PAS for the bubble sort routine given in Listing 6.3. However, before you enter that program, first put a copy of procedure Swap (Listing 6.2) at the beginning of the file. (If you are working with the Turbo Pascal editor, the command for this is Ctrl-KR.) Then add the lines given in Listing 6.3. Make a second, working copy of this file by giving the command

        COPY  SORTB.PAS  SORT.*

### Testing the Bubble Sort

Test our bubble sort routine by running the TESTSORT program. When you run the program, it asks how many numbers you want to sort. Select a number that is suitable for our timing test. Sorting speed is related to the speed of your computer. For example, a 12-MHz AT-compatible computer will run about ten times faster than an ordinary PC. Therefore, you will have to experiment to find an appropriate sample size. Start with 100 for a regular PC and 500 for an AT-compatible computer.

While you run the program, record the time difference between the three pairs of beeps. We will compare the times with those of the other sorting routines given in this chapter.

Next we will study a more sophisticated and slightly more difficult sorting routine.

# The Shell Sort

The major disadvantage of the bubble sort method is that it often makes more comparisons and more interchanges than are necessary.

```
PROCEDURE ( bubble ) Sort
        (VAR  A: Ary;
              N: Integer);
VAR
  I,J : Integer;
  Hold: Real;

BEGIN  ( procedure Sort )
  FOR I := 1 TO N-1 DO
    FOR J := I + 1 TO N DO
      IF A[I] > A[J] THEN
        Swap(A[I], A[J])
END; ( procedure Sort )
```

**Listing 6.3:** A bubble sort routine.

This method may move an item from one end of the array to the other. Then, a little later, it might move that item back to almost the same place where it started.

The *Shell-Metzner sort* (referred to as the Shell sort) is generally more efficient than the bubble sort. This method begins by making comparisons over long distances. It compares the first item in the array with one in the middle, rather than with the one right next to it. For short lists of fewer than a dozen items, the Shell and bubble sorts are comparable in speed. However, the Shell sort is noticeably faster than the bubble sort when the number of items exceeds about 50.

Let us compare the bubble and shell sorts by using our TESTSORT program.

## Programming the Shell Sort

Create another file named SORTS.PAS for the Shell sort routine. As before, first copy the procedure Swap, given in Listing 6.2, then add the lines shown in Listing 6.4. Make a working copy using the command

```
COPY  SORTS.PAS  SORT.*
```

## Testing the Shell Sort

Test the Shell sort routine by running the TESTSORT program again. Record the time difference between the beeps at the beginning and end of each step.

For a list of 800 items, the Shell sort is faster than the bubble sort by a factor of six. You will also notice that the Shell sort, like the bubble

```
PROCEDURE { shell } Sort
          (VAR   A: Ary;
                 N: Integer);
  { Shell-Metzner Sort }
  { Adapted from Programming in Pascal,
    P. Grogono, Addison-Wesley, 1980 }

VAR
  Done: Boolean;
  Jump, I, J: Integer;
```

**Listing 6.4:** A Shell sort routine.

```
BEGIN
   Jump := N;
   WHILE Jump > 1 DO
      BEGIN
         Jump := Jump DIV 2;
         REPEAT
            Done := True;
            FOR J := 1 TO N - Jump DO
               BEGIN
                  I := J + Jump;
                  IF A[J] > A[I] THEN
                     BEGIN
                        Swap(A[J], A[I]);
                        Done := False
                     END { if }
               END        { FOR }
         UNTIL Done
      END                 { WHILE }
END;                      { Sort }
```

**Listing 6.4:** A Shell Sort routine (continued).

sort, runs much faster with sorted data than it does with unsorted data.

In the next section, we will discuss our last and most complex sorting routine, which uses a recursive sorting procedure.

## The Quick Sort

We have seen that the Shell sort is a bit more complicated than the bubble sort, but it can sort much more quickly. Both of these algorithms can readily be programmed in other higher level languages, such as BASIC or Fortran. The third sorting algorithm we will consider is known as the *quick sort*. This method is more complicated than the Shell algorithm, but it is generally faster than the bubble or Shell sorts. However, if the original data are already sorted, or nearly sorted, the Shell sort can be much faster. A quick sort routine takes almost as long to run on a sorted array as on an unsorted one.

The bubble sort begins by comparing elements that are side by side. The Shell sort begins by comparing the initial element to one at the middle of the array. The quick sort begins by comparing the elements at the opposite ends of the array. Thus, this method can make the initial interchanges over large distances.

## Programming the Quick Sort

We will program a *recursive* version of the quick sort; that is, it contains a procedure that calls itself. A disadvantage of this algorithm is that it cannot be directly converted into BASIC or Fortran because of the recursive call. Furthermore, recursive routines can consume large amounts of memory.

Our quick sort routine uses the procedure Partit, which is repeatedly called to operate on various portions of the array. This procedure divides a given portion into two parts, and then it rearranges the elements so that all those on the left side are smaller than all those on the right. Next it divides the two new sections into two subsections, and the process is repeated. Partitioning continues until there are many sets containing one element each, at which point the array is sorted.

Create a file named SORTQ.PAS, and put a copy of procedure Swap from Listing 6.2 at the beginning. Then add the lines shown in Listing 6.5.

## Testing the Quick Sort

Test the quick sort routine by running the TESTSORT program a final time. Again, record the time difference between beeps. You will

```
PROCEDURE { quick } Sort
       (VAR X: Ary;
            N: Integer);
{ a recursive sorting routine }
{ Adapted from The design of Well-
     Structured and Correct Programs,
     S. Alagic, Springer-Verlag, 1978 }

PROCEDURE Qsort
       (VAR X: Ary;
           M, N: Integer);
VAR
  I, J: Integer;

PROCEDURE Partit
       (VAR    A: Ary;
           VAR I, J: Integer;
         Left, Right: Integer);
VAR
  Pivot: Sort_Var;
```

**Listing 6.5:** A recursive quick sort.

```
BEGIN
  Pivot := A[(Left + Right) DIV 2];
  I := Left;
  J := Right;
  WHILE I <= J DO
    BEGIN
      WHILE A[I] < Pivot DO
        I := I + 1;
      WHILE Pivot < A[J] DO
        J := J - 1;
      IF I <= J THEN
        BEGIN
          Swap(A[I], A[J]);
          I := I + 1;
          J := J - 1
        END
    END    { WHILE }
END;        { Partit }

BEGIN     { Qsort }
  IF M < N THEN
    BEGIN
      Partit(X, I, J, M, N); { divide in two }
      Qsort(X, M, J);  { Sort left part   }
      Qsort(X, I, N)   { Sort right part }
    END
END;        { Qsort }

BEGIN { Sort }
  Qsort(X, 1, N)
END;    { Sort }
```

**Listing 6.5:** A recursive quick sort (continued).

find that this method is the quickest for the unsorted data, but not for the sorted array.

You will want to keep versions of the Shell sort and quick sort routines. Both will prove useful. The Shell sort is preferred for smaller lists and for data that are almost sorted. The quick sort is better for large collections.

Now, let us return to our curve-fitting program, which is what originally led us into this discussion of sorting routines.

## *Adding a Sorting Routine to the Curve-Fitting Program*

The next step is to incorporate a sorting routine into the curve-fitting program that we wrote in Chapter 5. We will use the simpler Shell sort because we do not have very many data points. If you plan to work with larger collections of data, then you should add the quick sort routine instead. We need to make changes to both the curve-fitting program (Listing 5.8) and the sorting routine.

### *Revising the Curve-Fitting Program*

Make a copy of Listing 5.8, and give it the name CFIT6.PAS. We will edit the new file to make one change and add three new lines.

Near the beginning of the program, just after the TYPE heading, add the line

```
Sort_Var = Real;
```

Next change the following line to

```
Ary = ARRAY[1..Max] OF Sort_Var;
```

That is, change the word Real to Sort_Var. This region now looks like this:

```
TYPE
      Sort_Var = Real;
      Ary = ARRAY[1..Max] OF Sort_Var;
```

Go to the end of the program, and add the line

```
{$I SORT2.PAS}
```

just before the line

```
{$I PLOT.PAS}
```

For the final addition, go to the main program and, just after the line

```
Get_Data(X, Y, N);
```

add the line

```
Sort2(X, Y, N);
```

to call the sorting routine.

### *Revising the Shell Sort Routine*

The remaining changes are made to the sorting routine. Make a copy of the Shell sort procedure shown in Listing 6.4 with the command

```
COPY  SORTS.PAS  SORT2.*
```

We will edit the new SORT2 file to make three changes.

First, change the procedure name from Sort to Sort2. Next add the array Y to the second line of the heading just after symbol A. The revised beginning looks like this:

```
PROCEDURE { shell } Sort2
        (VAR A, Y : Ary;
                N : Integer);
```

Then locate the line

```
Swap(A[J], A[I]);
```

and add the line

```
Swap(Y[J], Y[I]);
```

right after it. Now, whenever a pair of X(A) values are interchanged, the corresponding Y values will also be interchanged.

If you choose to use the quick sort version instead, add the array Y to the heading like this:

```
PROCEDURE { quick } Sort
        (VAR X, Y : Ary;
                N : Integer);
```

Then, near the end of procedure Partit, immediately after the

```
Swap(A[I], A[J]);
```

add the line

```
Swap(Y[I], Y[J]);
```

## Running the Revised Curve-Fitting Program

Rerun the curve-fitting program after you have incorporated the sorting routine. Verify that the elements of array X are sorted in increasing order. Check that array Y has been rearranged along with X. The slope of the data displayed on the plot should now be in the opposite direction.

In the next section, we will develop a program for sorting data stored on disk.

# Sorting Data Stored on Disk

We have developed several routines for sorting real numbers created by a random-number generator. However, it is more likely that you will work with data stored on disk. Furthermore, the information may be strings of text made up of alphabetic and numeric characters, such as names and addresses. (Lines of alphabetic information are called records.) We will now make a new version of our sorting program that can sort alphabetic records stored on disk.

## Programming the Record Sorting Routine

Create a new file named SORTREC.PAS, and enter the lines shown in Listing 6.6. You can use any one of the sorting routines that we developed earlier in the chapter. Copy the routine that you choose in a file named SORT.PAS.

Next we need some records to sort. Create a working data file called STRING. Write the following lines into this file;

First line
Second line
Third line
Fourth line
Fifth line
Sixth line
Seventh line
Eighth line
Ninth line
Tenth line

```
PROGRAM SortRec;
{ Turbo Pascal program to sort alphabetic
  records stored on disk }

CONST
  Max = 200;

TYPE
  Sort_Var = STRING[120];
  Ary = ARRAY[1..Max] OF Sort_Var;
```

**Listing 6.6:** Program to sort alphabetic data on disk.

```
VAR
  X: Ary;
  N, I: Integer;
  Error: Boolean;
  Filename: STRING[14];
  Filvar1,Filvar2: Text;

PROCEDURE Get_Data;
{ read records from disk }

BEGIN {Get_Data}
   WriteLn;
   WriteLn(' Program to sort text files on disk ');
   REPEAT
      Write(' File to sort: ');
      ReadLn(Filename);
      Assign(Filvar1,Filename);
      {$I-}  {Turn off error checking}
      Reset(Filvar1);
      {$I+}  {Turn it back on}
   UNTIL IOresult = 0;
   N := 0;
   REPEAT
      N := N + 1;
      ReadLn(Filvar1,X[N])
   UNTIL Eof(Filvar1)
END; {Get_Data}

PROCEDURE Make_Fil;
{ Write records to disk }

BEGIN {Make_Fil}
  WriteLn;
  Write(' New file name: ');
  ReadLn(Filename);
  Assign(Filvar2,Filename);
  Rewrite(Filvar2);
  FOR I:= 1 TO N DO
     WriteLn(Filvar2,X[I]);
  Close(Filvar2);
END; {Make_Fil}

{$I SORT.PAS} {Listing 6.3, 6.4, or 6.5}

BEGIN  { main program }
   WriteLn;
   Get_Data;
   Sort(X, N);
   Make_Fil
END.
```

**Listing 6.6:** Program to sort alphabetic data on disk (continued).

### *Running the Record Sorting Program*

Run the new program. When the message

File to sort:

appears on the screen, enter the name of your data file:

STRING

and press the Enter key. The program will read the lines of this file into memory and sort them in alphabetic order. When the message

New file name:

appears, enter the name

SORTED

and press Enter.

Leave Turbo Pascal, and examine the new file by entering the command

TYPE SORTED

You will see that these ten lines have been placed in alphabetic order.

This very simple program can be the starting point for a more sophisticated version. For example, the program should check to see that a file with the new name does not already exist. If it does, the program should display an error message. You could also incorporate provisions for renaming the original file or requesting another name for the file.

## *Summary*

We now have two different sorting routines. We used these routines for sorting real numbers created by a random-number generator. We also incorporated a variation of one sorting program into the curve-fitting program that we developed in the previous chapter. Finally, we wrote a program to sort text files stored on disk.

Now that our curve-fitting program can handle real experimental data, we are ready to develop programs that can handle more complex equations, such as those described in Chapter 7.

# *Exercises*

**6-1.** Incorporate into the Shell sort procedure a Boolean variable called DFlag. This will be a direction flag set by the calling program. If the value is True, sorting will occur as usual. If the value of the flag is False, the data will be sorted in descending order. Make DFlag the last parameter of the procedure. Preset the flag to True in the calling program so that sorting will occur in the usual way. Ask the user if the data are to be sorted in reverse order. If the answer is positive, change the flag to False. In the sorting routine, there will be something like:

```
IF (DFlag AND (A[J] > A[I])
OR (not DFlag AND (A[J] < A[I]))) THEN . . .
```

**6-2.** Change one of the sorting routines so that it will sort strings obtained from the procedure Get_Data.

**6-3** Change one of the sorting routines so that it will sort strings obtained from the keyboard.

**6-4.** Alter the program given in Problem 6-3 so that the data are sorted on an interior field. Make columns 11 through 20 of the data correspond to a city. Then use these columns as the basis for sorting.

# *Generalized Least-Squares Curve Fitting*

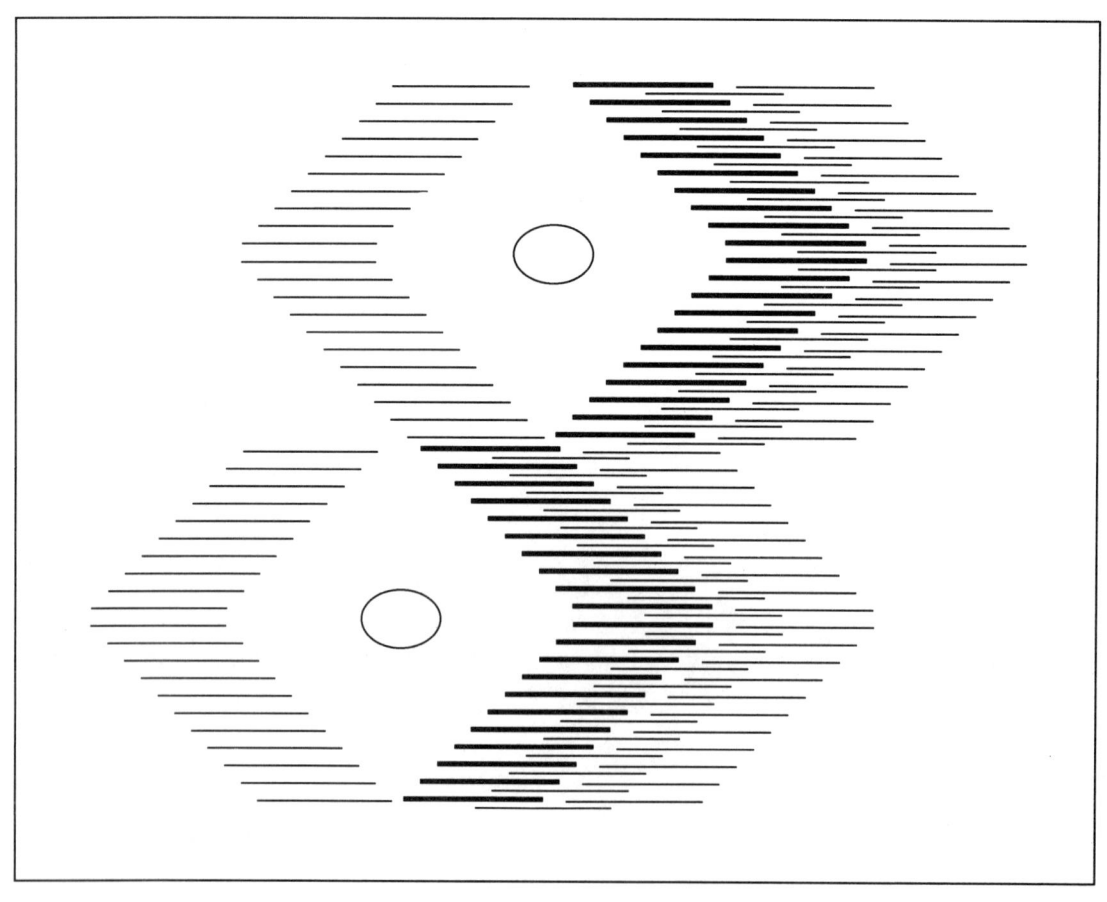

# Chapter 7

In Chapter 5, we developed a least-squares curve-fitting program. Now we will generalize that program so that it will be more useful in a wider variety of experimental situations. We will develop several different least-squares curve-fitting programs. The first program will generate a parabolic curve, that is, a second-order polynomial equation. From that point, we will learn how to fit data to both higher order polynomials and to nonpolynomial equations. The only restriction will be that the unknown coefficients to our approximating equation must be linear.

The key to our approach will be the use of a data vector and a data matrix. Using this method, we will be able to specify the order of a polynomial equation by entering a number from the keyboard while the program is running. At the end of the chapter, we will write curve-fitting programs for some real experimental data, including those for heat capacity, vapor pressure, and properties of superheated steam. Our final program uses a three-variable equation of state and includes a nonlinear coefficient that you specify.

# A Parabolic Curve Fit

The least-squares curve-fitting program that we developed in Chapter 5 calculated the coefficients A and B for the expression

$$Y = A + BX \tag{1}$$

While Equation 1 is used for many curve-fitting problems, sometimes a different equation is more appropriate.

In this chapter, we will develop a curve-fitting program for equations that are linear in the unknown coefficients. Thus, we will consider equations such as

$$Y = A + BX + CX^2$$

$$Y = A + \frac{B}{X} + CZ$$

$$\text{Ln } Y = \frac{AX}{Z} + Be^x$$

because the coefficients A, B, and C are linear. We will not consider an equation such as

$$Y = A + Be^{Cx}$$

because the coefficient C is not linear.

We will use an approach similar to the one that we used in Chapter 5. Consider, for example, the parabolic equation (a second-order polynomial)

$$Y = A + BX + CX^2 \tag{2}$$

We define the residuals to be

$$r = A + BX + CX^2 - Y$$

The residuals are squared and then summed. As in Chapter 5, to minimize this quantity, we take the derivative with respect to each variable (A, B, and C in this example) and set the resulting equations to zero. For the parabolic curve fit, there will be three equations—one for each of the three variables. The resulting equations are

$$AN + B\sum X + C\sum X^2 = \sum Y \tag{3}$$

$$A\sum X + B\sum X^2 + C\sum X^3 = \sum XY \tag{4}$$

$$A\sum X^2 + B\sum X^3 + C\sum X^4 = \sum X^2Y \tag{5}$$

## Programming
## a Least-Squares Curve Fit for a Parabola

We can obtain the solution to the parabolic least-squares fit by solving Equations 3, 4, and 5 simultaneously. The program shown in Listing 7.1 finds the solution to these equations by using Cramer's rule. The determinants of four 3-by-3 matrices are solved with this approach. We use function Deter, from Listing 4.1, for this purpose.

In the programs of the previous chapters, the unknown coefficients were represented by A, B, and C. However, in this chapter, we will use the vector Coef for this purpose. Thus, the constant term A now corresponds to the first element of the vector, Coef[1]. Similarly, the coefficients B and C correspond to Coef[2] and Coef[3], respectively.

Notice that Listing 7.1 uses procedure Plot, from Chapter 5. As before, it is referenced through an INCLUDE directive rather than being included with the main program. Procedure Plot requires the independent-variable array to be arranged in increasing or decreasing order. The data provided by procedure Get_Data are arranged in this manner. If you want to substitute other data that are not sorted, you can include one of the sorting routines discussed in Chapter 6.

```
PROGRAM Least1;
{ Turbo Pascal program to perform a
  linear least-squares fit
  using a parabolic curve.
  Separate procedure Plot needed }
{USES Crt;}   { Use for versions 4 and later }

CONST
  Maxr = 20;
  Maxc = 3;

TYPE
  Ary   = ARRAY[1..Maxr] OF Real;
  Arys  = ARRAY[1..Maxc] OF Real;
  Ary2s = ARRAY[1..Maxc, 1..Maxc] OF Real;

VAR
  X, Y, Y_Calc: Ary;
          Coef: Arys;
    Nrow, Ncol: Integer;
    Correl_Coef: Real;

{$I PLOT.PAS } {Listing 5.2}
```

**Listing 7.1:** A parabolic least-squares fit.

```
PROCEDURE Get_Data(VAR X, Y: Ary;
                   VAR Nrow: Integer);
{ Get values for Nrow and arrays X, Y }

VAR
  I: Integer;

BEGIN
  Nrow := 9;
  FOR I := 1 TO Nrow DO X[I] := I;
  Y[1]:=  2.07; Y[2]:=  8.6;
  Y[3]:= 14.42; Y[4]:= 15.80;
  Y[5]:= 18.92; Y[6]:= 17.96;
  Y[7]:= 12.98; Y[8]:=  6.45;
  Y[9]:=  0.27
END; { procedure Get_Data }

PROCEDURE Solve(A: Ary2s;
                Y: Arys;
            VAR Coef: Arys;
                Nrow: Integer;
            VAR Error: Boolean);

VAR
     B: Ary2s;
  I, J: Integer;
   Det: Real;

FUNCTION Deter(A: Ary2s): Real;
{ calculate the determinant of a 3-by-3 matrix }

BEGIN   { function Deter }
  Deter := A[1,1] *(A[2,2]*A[3,3] - A[3,2]*A[2,3])
         - A[1,2] *(A[2,1]*A[3,3] - A[3,1]*A[2,3])
         + A[1,3] *(A[2,1]*A[3,2] - A[3,1]*A[2,2])
END; { function Deter }

PROCEDURE Setup(VAR  B: Ary2s;
                VAR  Coef: Arys;
                     J: Integer);
VAR
  I: Integer;

BEGIN   { Setup }
  FOR I := 1 TO Nrow DO
    BEGIN
      B[I,J] := Y[I];
      IF J > 1 THEN B[I,J-1] := A[I,J-1]
    END;
    Coef[J] := Deter(B) / Det
END; { Setup }
```

**Listing 7.1:** A parabolic least-squares fit (continued).

```
BEGIN  { procedure Solve }
  Error := False;
  FOR I := 1 TO Nrow DO
    FOR J := 1 TO Nrow DO
      B[I,J] := A[I,J];
  Det := Deter(B);
  IF  Det = 0.0 THEN
    BEGIN
      Error := True;
      WriteLn(' ERROR: matrix singular')
    END
  ELSE
    BEGIN
      Setup(B, Coef, 1);
      Setup(B, Coef, 2);
      Setup(B, Coef, 3)
    END  { else }
END; { procedure Solve }

PROCEDURE Linfit(X, Y: Ary;
            VAR Y_Calc: Ary;
            VAR Coef  : Arys;
                Nrow  : Integer;
            VAR Ncol  : Integer);
{ least-squares fit to a parabola }
{ Nrow sets of X and Y pairs of points }


VAR
  A: Ary2s;
  G: Arys;
  I: Integer;
  Error: Boolean;
  Sum_X, Sum_Y, Sum_XY, Sum_X2,
  Sum_Y2, Xi, Yi, SXY ,SXX, SYY,
  Sum_X3, Sum_X4, Sum_2Y, Denom,
  SRS, X2: Real;

BEGIN  { Linfit }
  Ncol := 3; { polynomial terms }
  Sum_X := 0;
  Sum_Y := 0;
  Sum_XY := 0;
  Sum_X2 := 0;
  Sum_Y2 := 0;
  Sum_X3 := 0;
  Sum_X4 := 0;
  Sum_2Y := 0;
  FOR I := 1 TO Nrow DO
    BEGIN
      Xi := X[I];
      Yi := Y[I];
      X2 := Xi*Xi;
      Sum_X := Sum_X + Xi;
      Sum_Y := Sum_Y + Yi;
```

**Listing 7.1:** A parabolic least-squares fit (continued).

```
              Sum_XY := Sum_XY + Xi * Yi;
              Sum_X2 := Sum_X2 + X2;
              Sum_Y2 := Sum_Y2 + Yi * Yi;
              Sum_X3 := Sum_X3 + Xi * X2;
              Sum_X4 := Sum_X4 + X2 * X2;
              Sum_2Y := Sum_2Y + X2 * Yi
        END;
     A[1,1] := Nrow;
     A[2,1] := Sum_X;   A[1,2] := Sum_X;
     A[3,1] := Sum_X2;  A[1,3] := Sum_X2;
     A[2,2] := Sum_X2;  A[3,2] := Sum_X3;
     A[2,3] := Sum_X3;  A[3,3] := Sum_X4;
     G[1] := Sum_Y;
     G[2] := Sum_XY;
     G[3] := Sum_2Y;
     Solve(A, G, Coef, Ncol, Error);
     IF Error THEN Halt;
     SRS := 0.0;
     FOR I := 1 TO Nrow DO
        BEGIN
           Y_Calc[I] :=
              Coef[1] + Coef[2] * X[I] + Coef[3] * Sqr(X[I]);
           SRS := SRS + Sqr(Y[I] - Y_Calc[I])
        END;
     Correl_Coef :=
        Sqrt(1.0 - SRS/(Sum_Y2 -Sqr(Sum_Y)/Nrow))
  END; { Linfit }

  PROCEDURE Write_Data;
  { print out the answers }

  VAR
     I: Integer;

  BEGIN
     WriteLn;
     WriteLn(' I        X          Y      Y CALC');
     FOR I := 1 TO Nrow DO
        WriteLn(I:3, X[I]:8:1, Y[I]:9:2, Y_Calc[I]:9:2);
     WriteLn; WriteLn(' Coefficients');
     FOR I := 1 TO Ncol DO
        WriteLn(Coef[I]:8:4);
     WriteLn;
     WriteLn(' Correlation coefficient is ',Correl_Coef:8:5)
  END; { Write_Data }

  BEGIN  { main program }
     Get_Data(X, Y, Nrow);
     Linfit(X, Y, Y_Calc, Coef, Nrow, Ncol);
     Write_Data;
     WriteLn(' Press any key for plot');
     REPEAT UNTIL KeyPressed;
     Plot(X, Y, Y_Calc, Nrow)
  END.
```

**Listing 7.1:** A parabolic least-squares fit (continued).

## *Running the Program*

Create a file named LEAST1.PAS, and type the program shown in Listing 7.1. This program calculates a parabolic least-squares fit to the data. When you run it, the results should look like Figure 7.1. The correlation coefficient is close to unity, showing that this equation produces a good fit to the data. Figure 7.2 shows the plot from the parabolic

```
   I     X       Y     Y CALC
   1    1.0    2.07     1.68
   2    2.0    8.60     9.02
   3    3.0   14.42    14.20
   4    4.0   15.80    17.21
   5    5.0   18.92    18.05
   6    6.0   17.96    16.73
   7    7.0   12.98    13.24
   8    8.0    6.45     7.58
   9    9.0    0.27    -0.24

Coefficients
-7.8267
10.5900
-1.0830

Correlation coefficient is  0.99155
Press any key for plot
```

**Figure 7.1:** Output from the parabolic curve-fitting program.

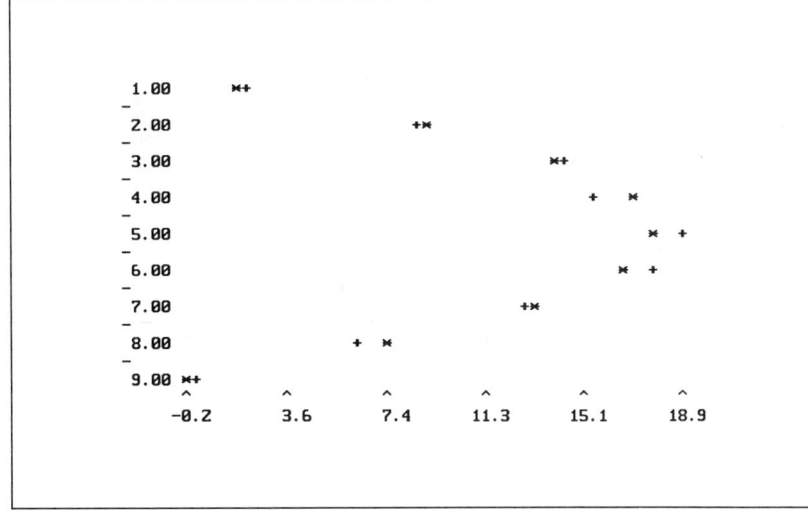

**Figure 7.2:** Plot from the parabolic curve-fitting program.

curve-fitting program. The resulting equation is

$$Y = -7.827 + 10.59X - 1.083X^2$$

Now let us consider polynomial equations with orders higher than 2 and nonpolynomial equations. We will see that the approach that we have been using to find the coefficients becomes less practical as the equation becomes more complex. Therefore, we will investigate another method.

## *Fitting Curves to Other Equations*

If we choose a higher order polynomial, there will be additional coefficients, resulting in more equations to be solved simultaneously. Equations 3, 4, and 5 can be extended easily. For example, to find the coefficients to the cubic equation

$$Y = A + BX + CX^2 + DX^3$$

we define the residuals as

$$r = A + BX + CX^2 + DX^3 - Y$$

The residuals are first squared and then summed. The derivative is taken with respect to each of the four variables A, B, C, and D. The resulting four equations are solved simultaneously.

$$AN + B\sum X + C\sum X^2 + D\sum X^3 = \sum Y$$

$$A\sum X + B\sum X^2 + C\sum X^3 + D\sum X^4 = \sum XY$$

$$A\sum X^2 + B\sum X^3 + C\sum X^4 + D\sum X^5 = \sum X^2 Y$$

$$A\sum X^3 + B\sum X^4 + C\sum X^5 + D\sum X^6 = \sum X^3 Y$$

Expressions that are not polynomials are treated similarly. For example, the general three-term equation is

$$Y = A + Bf(x) + Cg(x)$$

where f(x) and g(x) represent any function of X. The residuals, defined as

$$r = A + Bf(x) + Cg(x) - Y$$

are squared and then summed. The derivatives with respect to the three variables give the three equations

$$A\sum f(x) + B\sum f(x)^2 + C\sum f(x)g(x) = \sum f(x)Y$$

$$An + B\sum f(x) + C\sum g(x) = \sum Y$$

$$A\sum g(x) + B\sum f(x)g(x) + C\sum g(x)^2 = \sum g(x)Y$$

Although the above approach to curve fitting is correct, it is laborious. We need to make major alterations whenever we change the curve-fitting equation. For example, if the equation were chosen to be

$$Y = A + BX + C/X^2$$

then we would need to include statements in the program such as

```
Sum_X3 := Sum_X3 + 1/ Xi;
Sum_X4 := Sum_X4 + 1/(Xi*Xi);
```

for calculating the needed sums.

## A Direct Solution

A better way to determine the coefficients of the curve-fitting equation is to set up the data in a matrix and a vector. We then convert the data matrix and vector to a set of simultaneous equations that can be solved by using one of the methods that we discussed in Chapter 4. For example, suppose that we want a linear fit to the equation

$$Y = A + BX$$

for five sets of x-y data. The data vector would simply be the vector of Y values. The data matrix would look like this:

$$\begin{bmatrix} 1 & X_1 \\ 1 & X_2 \\ 1 & X_3 \\ 1 & X_4 \\ 1 & X_5 \end{bmatrix}$$

Each row of the data matrix corresponds to one of the data points, and each column corresponds to one term of the equation. As a result, the data matrix has five rows and two columns. Column 1 of the matrix has only the value of unity, since that is the corresponding function of X (that is, $X^0$) in the first term of the equation. Column 2 contains the values of X (that is, $X^1$) because that is the function of X in the second term of the equation.

The data matrix for a parabolic fit has three columns. The first two columns are the same as for the straight-line fit. The third column, however, contains the square of each X value. If, on the other hand, we choose an equation like

$$\ln P = A + \frac{B}{T} + C \ln t$$

then, the first column of the data matrix contains the value of 1, the second column has the reciprocal of the data, and the third column contains the logarithm. The data vector is this case has the logarithm of P.

We convert the rectangular data matrix to a square matrix by using a simple operation. We multiply the transpose of the data matrix by the matrix itself to produce the coefficient matrix. For the straight-line fit of five sets of x-y data, the operation is

$$\begin{bmatrix} 1 & 1 & 1 & 1 & 1 \\ X_1 & X_2 & X_3 & X_4 & X_5 \end{bmatrix} \begin{bmatrix} 1 & X_1 \\ 1 & X_2 \\ 1 & X_3 \\ 1 & X_4 \\ 1 & X_5 \end{bmatrix}$$

The result is a 2-by-2 matrix containing the required sums of X:

$$\begin{bmatrix} N & \Sigma X \\ \Sigma X & \Sigma X^2 \end{bmatrix}$$

The product of the data vector (considered a row vector) and the data matrix

$$\begin{bmatrix} Y_1 & Y_2 & Y_3 & Y_4 & Y_5 \end{bmatrix} \begin{bmatrix} 1 & X_1 \\ 1 & X_2 \\ 1 & X_3 \\ 1 & X_4 \\ 1 & X_5 \end{bmatrix}$$

gives the required constant vector of length 2:

$$\begin{bmatrix} \Sigma Y & \Sigma XY \end{bmatrix}$$

We will now examine a Turbo Pascal version of this approach. We will also want to incorporate into the program a general technique for calculating the standard error.

## Programming the Matrix Approach to Curve Fitting

Listing 7.2 gives a curve-fitting program that uses this matrix approach. The data matrix and vector are set up at the beginning of

```
PROGRAM Least2;
{ Turbo Pascal program to perform a
  linear least-squares fit
  with Gauss-Jordan elimination routine.
  Separate procedures needed:
        Gaussj, Plot, and Square }
{USES Crt;}  { Use for versions 4 and later }

CONST
  Maxr = 20;        { data points }
  Maxc = 4;         { polynomial terms }

TYPE
  Ary   = ARRAY[1..Maxr] OF Real;
  Arys  = ARRAY[1..Maxc] OF Real;
  Ary2  = ARRAY[1..Maxr, 1..Maxc] OF Real;
  Ary2s = ARRAY[1..Maxc, 1..Maxc] OF Real;

VAR
  X, Y, Y_Calc, Resid: Ary;
  Coef, Sig  : Arys;
  Nrow, Ncol : Integer;
  Correl_Coef: Real;

{$I SQUARE.PAS } {Listing 3.2}
{$I GAUSSJ.PAS } {Listing 4.4}
{$I PLOT.PAS }   {Listing 5.2}

PROCEDURE Get_Data
          (VAR X    : Ary; { independent variable }
           VAR Y    : Ary; { dependent variable }
           VAR Nrow: Integer); { length of vectors }

VAR
  I: Integer;

BEGIN
  Nrow := 9;
  FOR I := 1 TO Nrow DO X[I] := I;
  Y[1]:=  2.07; Y[2]:=   8.6;
  Y[3]:= 14.42; Y[4]:= 15.80;
  Y[5]:= 18.92; Y[6]:= 17.96;
  Y[7]:= 12.98; Y[8]:=  6.45;
  Y[9]:=  0.27
END;             { procedure Get_Data }

PROCEDURE Linfit(X,        { independent variable }
                 Y: Ary;   { dependent variable }
            VAR Y_Calc: Ary;  { calculated dependent variable }
            VAR Resid : Ary;  { array of residuals }
            VAR  Coef : Arys; { coefficients }
            VAR  Sig  : Arys; { errors in coefficients }
                 Nrow  : Integer;  { length of Ary }
            VAR  Ncol  : Integer); { number of terms }
```

**Listing 7.2:** A parabolic least-squares fit using Gauss-Jordan elimination.

```
{ least-squares fit to }
{ Nrow sets of X and Y pairs of points. }
{ Separate procedures needed:
  Square and Gaussj }

VAR
    Xmatr: Ary2;  { data matrix }
        A: Ary2s; { coefficient matrix }
        G: Arys;  { constant vector }
    Error: Boolean;
  I, J, Nm: Integer;
  Xi, Yi, Yc, SRS, SEE,
  Sum_Y, Sum_Y2: Real;

BEGIN             { procedure Linfit }
  Ncol := 3;      { number of terms }
  FOR I := 1 TO Nrow DO
    BEGIN   { Setup X matrix }
      Xi := X[I];
      Xmatr[I, 1] := 1.0;  { first column }
      Xmatr[I, 2] := Xi;   { second column }
      Xmatr[I, 3] := Xi*Xi { third column }
    END;
  Square(Xmatr, Y, A, G, Nrow, Ncol);
  Gaussj(A, G, Coef, Ncol, Error);
  Sum_Y := 0.0;
  Sum_Y2 := 0.0;
  SRS := 0.0;
  FOR I := 1 TO Nrow DO
    BEGIN
      Yi := Y[I];
      Yc := 0.0;
      FOR J := 1 TO Ncol DO
        Yc := Yc + Coef[J] * Xmatr[I, J];
      Y_Calc[I] := Yc;
      Resid[I] := Yc - Yi;
      SRS := SRS + Sqr(Resid[I]);
      Sum_Y := Sum_Y + Yi;
      Sum_Y2 := Sum_Y2 + Yi * Yi
    END;
  Correl_Coef :=
      Sqrt(1.0 - SRS /(Sum_Y2 - Sqr(Sum_Y) / Nrow));
  IF Nrow = Ncol THEN Nm := 1
  ELSE Nm := Nrow - Ncol;
  SEE := Sqrt(SRS / Nm);
  FOR I := 1 TO Ncol DO   { errors in solution }
    Sig[I] := SEE * Sqrt(A[I, I])
END;   { Linfit }

PROCEDURE Write_Data;
{ print out the answers }

VAR
  I: Integer;
```

**Listing 7.2:** A parabolic least-squares fit using Gauss-Jordan elimination (continued).

```
BEGIN
  WriteLn;
  WriteLn;
  WriteLn(' I      X       Y      Y CALC     RESID');
  FOR I := 1 TO Nrow DO
    WriteLn(I:3, X[I]:8:1, Y[I]:9:2,
      Y_Calc[I]:9:2, Resid[I]:9:2);
  WriteLn;
  WriteLn('coefficients      errors');
  WriteLn(Coef[1]:12,'   ', Sig[1]:12,' Constant term');
  FOR I := 2 TO Ncol DO
    WriteLn
      (Coef[I]:12,'   ', Sig[I]:12);   { other terms }
  WriteLn;
  WriteLn
    (' Correlation coefficient is ', Correl_Coef:8:5)
END;      { Write_Data }

BEGIN              { main program }
  Get_Data(X, Y, Nrow);
  Linfit(X, Y, Y_Calc, Resid, Coef, Sig, Nrow, Ncol);
  Write_Data;
  WriteLn(' Press any key for plot');
  REPEAT UNTIL KeyPressed;
  Plot(X, Y, Y_Calc, Nrow)
END.
```

**Listing 7.2:** A parabolic least-squares fit using Gauss-Jordan elimination (continued).

procedure Linfit. Then, procedure Square, which we used in Chapter 3, converts the data matrix X and data vector Y into the square coefficient matrix A and the constant vector G. The operations are

$$X^TX = A \quad \text{and} \quad YX = G$$

The program multiplies the data vector by the data matrix to produce the constant vector. The solution vector

$$A^{-1}G = B$$

can be obtained by using any of the routines we developed in Chapter 4 for solving simultaneous equations.

The linear curve-fitting program developed in Chapter 5 presented the standard errors along with the corresponding elements of the solution to the approximating function. It obtained the standard errors from the SEE and the summation of X and $X^2$. For our parabolic curve-fitting program, we will use a more general technique.

The standard errors are readily obtained from the inverse of the coefficient matrix A. The value corresponding to the *ith* term of the approximating function is the product of the SEE and the square root of the *ith* term of the major diagonal of the inverse

$$\begin{bmatrix} A_{11} & & \\ & A_{22} & \\ & & A_{33} \end{bmatrix}^{-1}$$

The Turbo Pascal expression is

    Sig[I] := SEE * Sqrt(A[I,I])

In Chapter 4, we discussed several methods for solving simultaneous equations. However, only the Gauss-Jordan elimination method generates the inverse of the coefficient matrix. Since we need this inverse to determine the errors in the corresponding elements of the solution vector, we will use the Gauss-Jordan elimination method for the remaining curve-fitting programs in this chapter.

We can use the standard errors in the coefficients to determine the confidence intervals for the corresponding coefficients. In addition, the standard errors can alert us to the possibility of ill conditioning. Although ill-conditioned matrices (discussed in Chapter 4) are more likely to occur in the solution of simultaneous equations than in curve fitting, we still want to watch for such a problem. The standard errors are derived from the square roots of the diagonal elements of the inverted matrix. If the squares of the errors are many orders apart, then ill conditioning may be present. We will see a demonstration of ill conditioning later in the chapter.

The program shown in Listing 7.2 is similar to the one given in Listing 7.1, but it solves the parabolic equation with the more general matrix method. By using the INCLUDE directive, the program references three routines that we developed in earlier chapters: the matrix multiplication procedure Square (from Chapter 3), the Gauss-Jordan elimination procedure Gaussj (from Chapter 4), and procedure Plot (from Chapter 5).

## *Running the Parabolic Curve-Fitting Program*

Make a copy of the program in Listing 7.1, and give it the name LEAST2.PAS. Change the copy to look like the one given in Listing 7.2. Be sure that you have procedures Square, Gaussj, and Plot available

```
     I      X       Y      Y CALC    RESID
     1     1.0     2.07     1.68     -0.39
     2     2.0     8.60     9.02      0.42
     3     3.0    14.42    14.20     -0.22
     4     4.0    15.80    17.21      1.41
     5     5.0    18.92    18.05     -0.87
     6     6.0    17.96    16.73     -1.23
     7     7.0    12.98    13.24      0.26
     8     8.0     6.45     7.58      1.13
     9     9.0     0.27    -0.24     -0.51

  coefficients       errors
  -7.8267E+000    1.29802E+000   Constant term
  1.05900E+001    5.96001E-001
  -1.0830E+000    5.81269E-002

  Correlation coefficient is    0.99155
```

**Figure 7.3:** Output from an alternative version of a parabolic least-squares curve-fitting program.

as separate files. Run the program, and compare the output with Figure 7.3. The results are the same as those given by the previous program, except that the new version also gives the residuals and errors.

In the next section, we will use the matrix approach to fit different orders of polynomial equations to any given set of data. To develop a sense of the full power of the revised program, we will run it several times on one set of data. By comparing the resulting plotted curves and correlation coefficients, we will find the best polynomial order to fit our data.

## *Adjusting the Order of the Polynomial*

One of the advantages of the new version of our curve-fitting program is that we can readily change both the number of rows and the number of columns, which correspond to the number of data points and the number of polynomial terms in the equation, respectively. Therefore, in a third version of the program, we will proceed one step further in generalization. From the keyboard, we will enter the order of the polynomial equation, which is a value one smaller than the number of terms in the equation. Make a copy of the program shown in Listing 7.2, alter it to look like the one in Listing 7.3, and give it the name LEAST3.PAS. You will need to revise the main program and procedure Linfit.

```
PROGRAM Least3;
{ Turbo Pascal program to perform a
  linear least-squares fit
  with Gauss-Jordan elimination routine.
  Separate procedures needed:
        Square, Gaussj, and Plot  }
{USES Crt;}  { Use for versions 4 and later }

CONST
  Maxr = 20;      { data points }
  Maxc = 4;       { polynomial terms }

TYPE
  Ary   = ARRAY[1..Maxr] OF Real;
  Arys  = ARRAY[1..Maxc] OF Real;
  Ary2  = ARRAY[1..Maxr, 1..Maxc] OF Real;
  Ary2s = ARRAY[1..Maxc, 1..Maxc] OF Real;

VAR
  X, Y, Y_Calc, Resid: Ary;
    Coef, Sig: Arys;
   Nrow, Ncol: Integer;
  Correl_Coef: Real;
         Done: Boolean;

{$I SQUARE.PAS } {Listing 3.2}
{$I GAUSSJ.PAS } {Listing 4.4}
{$I PLOT.PAS }   {Listing 5.2}

PROCEDURE Get_Data
        (VAR X   : Ary; { independent variable }
         VAR Y   : Ary; { dependent variable }
         VAR Nrow: Integer); { length of vectors }
VAR
  I: Integer;

BEGIN
  Nrow := 9;
  FOR I := 1 TO Nrow DO X[I] := I;
  Y[1]:=  2.07; Y[2]:=  8.6;
  Y[3]:= 14.42; Y[4]:= 15.80;
  Y[5]:= 18.92; Y[6]:= 17.96;
  Y[7]:= 12.98; Y[8]:=  6.45;
  Y[9]:=  0.27
END;  { procedure Get_Data }

PROCEDURE Linfit(X,       { independent variable }
                 Y: Ary;  { dependent variable }
         VAR Y_Calc: Ary; { calculated dependent variable }
         VAR Resid : Ary; { array of residuals }
         VAR Coef  : Arys; { coefficients }
         VAR Sig   : Arys; { errors in coefficients }
             Nrow : Integer; { length of Ary }
         VAR Ncol : Integer); { number of terms }
```

**Listing 7.3:** Inputting the polynomial order from the keyboard.

```
{ least-squares fit to }
{ Nrow sets of X and Y pairs of points. }
{ Separate procedures needed:
  Square and Gaussj }

VAR
    Xmatr: Ary2;  { data matrix }
       A: Ary2s; { coefficient matrix }
       G: Arys;  { constant vector }
    Error: Boolean;
  I, J, Nm: Integer;
  Xi, Yi, Yc, SRS, SEE,
  Sum_Y, Sum_Y2: Real;

BEGIN                { procedure Linfit }
  FOR I := 1 TO Nrow DO
    BEGIN    { Setup X matrix }
      Xi := X[I];
      Xmatr[I, 1] := 1.0;       { first column }
      FOR J := 2 TO Ncol DO    { other columns }
        Xmatr[I, J] := Xmatr[I,J-1] * Xi
    END;
  Square(Xmatr, Y, A, G, Nrow, Ncol);
  Gaussj(A, G, Coef, Ncol, Error);
  Sum_Y := 0.0;
  Sum_Y2 := 0.0;
  SRS := 0.0;
  FOR I := 1 TO Nrow DO
    BEGIN
      Yi := Y[I];
      Yc := 0.0;
      FOR J := 1 TO Ncol DO
        Yc := Yc + Coef[J] * Xmatr[I, J];
      Y_Calc[I] := Yc;
      Resid[I] := Yc - Yi;
      SRS := SRS + Sqr(Resid[I]);
      Sum_Y := Sum_Y + Yi;
      Sum_Y2 := Sum_Y2 + Yi * Yi
    END;
  Correl_Coef :=
      Sqrt(1.0 - SRS /(Sum_Y2 - Sqr(Sum_Y) / Nrow));
  IF Nrow = Ncol THEN Nm := 1
  ELSE Nm := Nrow - Ncol;
  SEE := Sqrt(SRS / Nm);
  FOR I := 1 TO Ncol DO   { errors in solution }
    Sig[I] := SEE * Sqrt(A[I, I])
END;  { Linfit }

PROCEDURE Write_Data;
{ print out the answers }

VAR
  I: Integer;
```

**Listing 7.3:** Inputting the polynomial order from the keyboard (continued).

```
BEGIN
  WriteLn;
  WriteLn;
  WriteLn(' I       X       Y       Y CALC      RESID');
  FOR I := 1 TO Nrow DO
    WriteLn(I:3, X[I]:8:1, Y[I]:9:2,
      Y_Calc[I]:9:2, Resid[I]:9:2);
  WriteLn;
  WriteLn('coefficients     errors');
  WriteLn(Coef[1]:12,'    ', Sig[1]:12,'  Constant term');
  FOR I := 2 TO Ncol DO
    WriteLn
      (Coef[I]:12,'    ', Sig[I]:12);    { other terms }
  WriteLn;
  WriteLn
    (' Correlation coefficient is ', Correl_Coef:8:5)
END;   { Write_Data }

BEGIN   { main program }
  Done := False;
  WriteLn;
  Get_Data(X, Y, Nrow);
  REPEAT
    REPEAT
      Write(' Order of polynomial fit? ');
      ReadLn(Ncol)
    UNTIL Ncol < Maxc;
    IF Ncol < 1 THEN
      Done := True   { quit if Ncol <1 }
    ELSE
      BEGIN
        Ncol := Ncol +1; { order is one less }
        Linfit(X, Y, Y_Calc, Resid, Coef, Sig, Nrow, Ncol);
        Write_Data;
        WriteLn(' Press any key for plot');
        REPEAT UNTIL KeyPressed;
        Plot(X, Y, Y_Calc, Nrow)
      END   { ELSE }
  UNTIL Done
END.
```

**Listing 7.3:** Inputting the polynomial order from the keyboard (continued).

## Comparing Runs
## Using Different Polynomial Orders

Run the program, and enter a value of 2 for the polynomial order. The results should again be the same as the ones obtained by the two previous versions. This time, however, the program will cycle and ask for the polynomial order again. Give a value of 1 the second time. The results (shown in Figure 7.4) are the best straight line through the curved set of data. Figure 7.5 shows the plot of this straight-line fit.

The two coefficients represent the equation

$$Y = 12.028 - 0.240X$$

Notice that about one-half of the data points are on one side of the straight line and one-half are on the other. The straight line goes through the points, as best it can. The correlation coefficient, however, is only 0.096. This relatively small value shows that the straight-line fit is not very good.

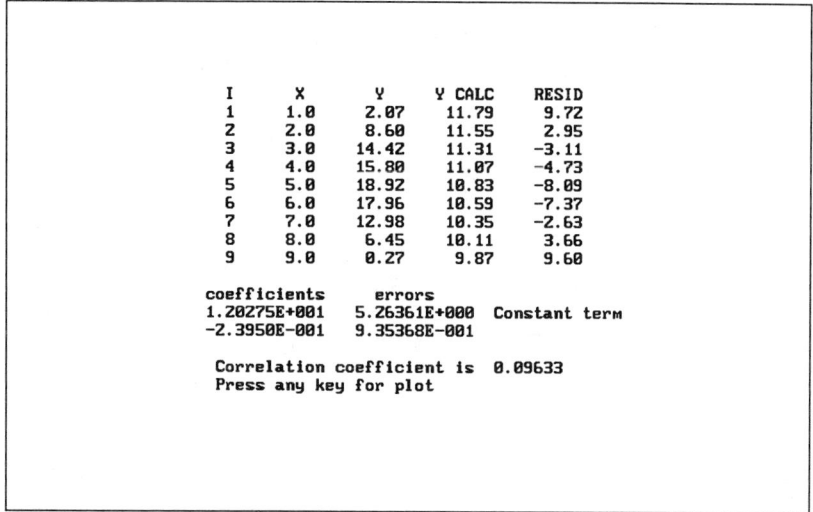

**Figure 7.4:** Output from the program using an adjusted polynomial order.

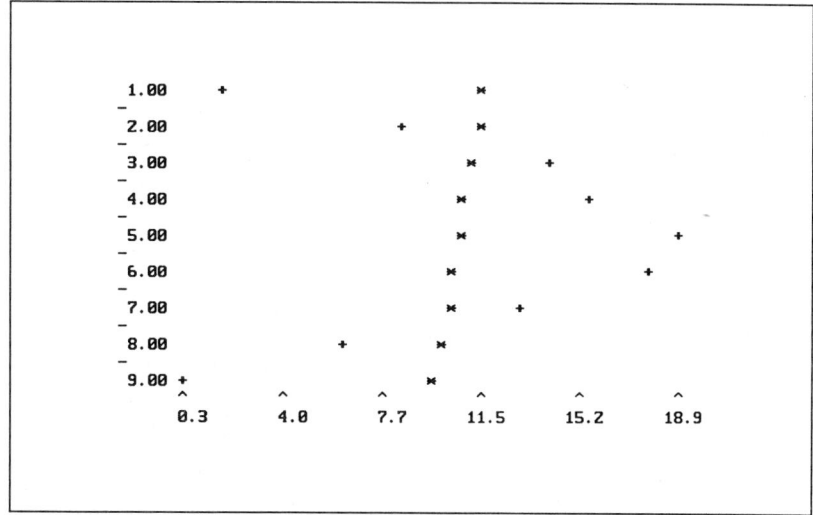

**Figure 7.5:** Plot of the straight-line fit to parabolic data.

Give an order of 3 for the third cycle. This will produce a fit corresponding to the cubic equation

$$Y = -7.2 + 10.0X - 0.946X^2 - 0.00915X^3$$

From the value of the correlation coefficient, we can see that the resulting curve describes the data no better than the parabolic fit. Since you should always use the lowest order equation that reasonably fits the data, the parabola is the better choice this time.

In this version of the program, we are limited to a polynomial order of 3 because we set the maximum number of matrix columns, the constant Maxc, to 4. If you need to use a higher order polynomial, change Maxc to a correspondingly larger value.

You can terminate the program by entering a polynomial order of 0, or simply by pressing Ctrl-C. You may want to alter the program so that it asks for the number of terms, rather than the polynomial order. Then you should remove the statement

```
Ncol := Ncol + 1; { order is one less }
```

from the main program.

In the next three sections, we will look at experimental applications of curve fitting to solve problems involving the heat capacity of oxygen, the vapor pressure of liquid lead, and the properties of superheated steam.


## *The Heat-Capacity Equation*

Heat capacity is a measure of how much the temperature of a mass will increase when a given amount of heat is added. Experimentally determined data are commonly fitted to the equation

$$C_p = A + BT + C/T^2$$

where $C_p$ is the heat capacity in units of energy per degree; T is the absolute temperature; and A, B, and C are the coefficients to be determined.

Make a copy of the curve-fitting program given in Listing 7.2 (not the one shown in Listing 7.3, which is designed for adjusting the polynomial order), and give the copy the name LEAST4.PAS. Alter procedures Get_Data and Linfit to look like the versions shown in Listing 7.4. As before, procedure Linfit fills the first column of the data matrix with the value of unity and the second column with the independent

```
PROCEDURE Get_Data
         (VAR T    : Ary; { independent variable }
          VAR Cp   : Ary; { dependent variable }
          VAR Nrow: Integer); { length of vectors }
VAR
  I: Integer;

BEGIN
  Nrow := 10;
  FOR I := 1 TO Nrow DO
    T[I] := (I + 2)* 100;
  Cp[1]:=  7.02; Cp[2]:=  7.20;
  Cp[3]:=  7.43; Cp[4]:=  7.67;
  Cp[5]:=  7.88; Cp[6]:=  8.06;
  Cp[7]:=  8.21; Cp[8]:=  8.34;
  Cp[9]:=  8.44; Cp[10]:= 8.53
END; { procedure Get_Data }

PROCEDURE Linfit(X,          { independent variable }
             Y: Ary;     { dependent variable }
         VAR Y_Calc: Ary; { calculated dependent variable }
         VAR Resid : Ary;  { array of residuals }
         VAR Coef  : Arys; { coefficients }
         VAR Sig   : Arys; { errors in coefficients }
             Nrow  : Integer; { length of Ary }
         VAR Ncol  : Integer); { number of terms }

{ least-squares fit to }
{ Nrow sets of X and Y pairs of points. }
{ Separate procedures needed:
    Square and Gaussj }

VAR
     Xmatr: Ary2; { data matrix }
         A: Ary2s; { coefficient matrix }
         G: Arys;  { constant vector }
     Error: Boolean;
  I, J, Nm: Integer;
  Xi, Yi, Yc, SRS, SEE,
  Sum_Y, Sum_Y2: Real;

BEGIN         { procedure Linfit }
  Ncol := 3; { number of terms }
  FOR I := 1 TO Nrow DO
    BEGIN    { Setup X matrix }
      Xi := X[I];
      Xmatr[I, 1] := 1.0;  { first column }
      Xmatr[I, 2] := Xi;   { second column }
      Xmatr[I, 3] := 1.0/Sqr(Xi) { third column }
    END;

  Square(Xmatr, Y, A, G, Nrow, Ncol);
  Gaussj(A, G, Coef, Ncol, Error);
  Sum_Y := 0.0;
  Sum_Y2 := 0.0;
  SRS := 0.0;
```

**Listing 7.4:** Procedures Get_Data and Linfit for the heat-capacity equation.

```
FOR I := 1 TO Nrow DO
   BEGIN
      Yi := Y[I];
      Yc := 0.0;
      FOR J := 1 TO Ncol DO
         Yc := Yc + Coef[J] * Xmatr[I, J];
      Y_Calc[I] := Yc;
      Resid[I] := Yc - Yi;
      SRS := SRS + Sqr(Resid[I]);
      Sum_Y := Sum_Y + Yi;
      Sum_Y2 := Sum_Y2 + Yi * Yi
   END;
Correl_Coef :=
      Sqrt(1.0 - SRS /(Sum_Y2 - Sqr(Sum_Y) / Nrow));
IF Nrow = Ncol THEN Nm := 1
ELSE Nm := Nrow - Ncol;
SEE := Sqrt(SRS / Nm);
FOR I := 1 TO Ncol DO     { errors in solution }
   Sig[I] := SEE * Sqrt(A[I, I])
END; { Linfit }
```

**Listing 7.4:** Procedures Get_Data and Linfit for the heat-capacity equation (continued).

```
I      X        Y      Y CALC    RESID
1     300.0    7.02     6.97     -0.05
2     400.0    7.20     7.27      0.07
3     500.0    7.43     7.49      0.06
4     600.0    7.67     7.67      0.00
5     700.0    7.88     7.84     -0.04
6     800.0    8.06     8.00     -0.06
7     900.0    8.21     8.16     -0.05
8    1000.0    8.34     8.31     -0.03
9    1100.0    8.44     8.46      0.02
10   1200.0    8.53     8.60      0.07

coefficients      errors
6.90485E+000    1.29216E-001  Constant term
1.43449E-003    1.26803E-004
-3.2610E+004    1.16749E+004

Correlation coefficient is  0.99482
Press any key for plot
```

**Figure 7.6:** Output from the heat capacity of oxygen calculation.

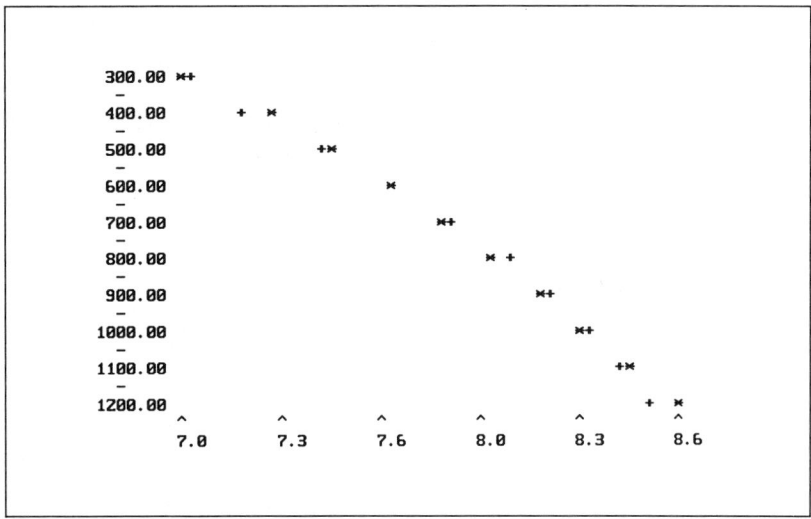

**Figure 7.7:** Plot of the heat capacity of oxygen.

variable (temperature here). However, now the third column contains the reciprocal of the temperature squared.

The data represent the heat capacity of oxygen over the temperature range of 300 to 1200 kelvins. Notice that we changed X and Y in procedure Get_Data to T and CP. Because these names are local to procedure Get_Data, we can change them to whatever is convenient.

When you run the program, the results will look like Figure 7.6. The plot of the heat capacity of oxygen is shown in Figure 7.7. The resulting equation is

$$C_p = 6.9 + 0.00143T - 32610/T^2$$

## *The Vapor Pressure Equation*

When a gas or vapor is in equilibrium with its own liquid or solid, we say that it is saturated. For pure materials, the saturation pressure is a single-valued function of the temperature. An equation commonly used to express the relationship between the saturation pressure and the saturation temperature is

$$\ln P = A + \frac{B}{T} + C \ln T$$

In this equation, P is the pressure; T is the absolute temperature; and A, B, and C are the coefficients.

Make a copy of the program shown in Listing 7.4 using the name LEAST5.PAS. Alter procedures Get_Data and Linfit to look like the versions shown in Listing 7.5. Procedure Linfit sets the first column of the matrix to one, as usual. Column 2 is then filled with the reciprocal of the independent variable, the temperature. The logarithm of the temperature goes in column 3.

Run the new program. The results should look like Figure 7.8. The data represent the vapor pressure of liquid lead over the temperature range of 700 to 1600 kelvins. The corresponding equation is

$$\ln P = 18.1 - \frac{23170}{T} - 0.884 \ln T$$

when the pressure is in atmospheres and the natural logarithm is used. The correlation coefficient of 1.00000 shows a very good fit.

```
PROCEDURE Get_Data
        (VAR T   : Ary; { independent variable }
         VAR P   : Ary; { dependent variable }
         VAR Nrow: Integer); { length of vectors }
VAR
   I: Integer;

BEGIN { Get_Data }
   Nrow := 10;
   FOR I := 1 TO Nrow DO
     T[I] := (I + 6.0)* 100.0;
   P[1]:=   1.0E-9;    P[2]:=   5.598E-8;
   P[3]:=   1.234E-6;  P[4]:=   1.507E-5;
   P[5]:=   1.138E-4;  P[6]:=   6.067E-4;
   P[7]:=   2.512E-3;  P[8]:=   8.337E-3;
   P[9]:=   2.371E-2;  P[10]:=  5.875E-2;
   FOR I := 1 TO Nrow DO
     P[I] := Ln(P[I])   { take log of P }
END; { procedure Get_Data }

PROCEDURE Linfit(X,        { independent variable }
                 Y: Ary;   { dependent variable }
         VAR Y_Calc: Ary;  { calculated dependent variable }
         VAR Resid : Ary;  { array of residuals }
         VAR  Coef : Arys; { coefficients }
         VAR   Sig : Arys; { errors in coefficients }
              Nrow : Integer;  { length of Ary }
         VAR  Ncol : Integer); { number of terms }
```

**Listing 7.5:** Procedure Get_Data and Linfit for the vapor pressure equation.

```
{ least-squares fit to }
{ Nrow sets of X and Y pairs of points. }
{ Separate procedures needed:
   Square and Gaussj }

VAR
    Xmatr: Ary2;  { data matrix }
        A: Ary2s; { coefficient matrix }
        G: Arys;  { constant vector }
    Error: Boolean;
 I, J, Nm: Integer;
Xi, Yi, Yc, SRS, SEE,
Sum_Y, Sum_Y2: Real;

BEGIN          { procedure Linfit }
  Ncol := 3; { number of terms }
  FOR I := 1 TO Nrow DO
    BEGIN   { Setup X matrix }
      Xi := X[I];
      Xmatr[I, 1] := 1.0;       { first column }
      Xmatr[I, 2] := 1.0/Xi; { second column }
      Xmatr[I, 3] := Ln(Xi)  { third column }
    END;

  Square(Xmatr, Y, A, G, Nrow, Ncol);
  Gaussj(A, G, Coef, Ncol, Error);
  Sum_Y := 0.0;
  Sum_Y2 := 0.0;
  SRS := 0.0;
  FOR I := 1 TO Nrow DO
    BEGIN
      Yi := Y[I];
      Yc := 0.0;
      FOR J := 1 TO Ncol DO
        Yc := Yc + Coef[J] * Xmatr[I, J];
      Y_Calc[I] := Yc;
      Resid[I] := Yc - Yi;
      SRS := SRS + Sqr(Resid[I]);
      Sum_Y := Sum_Y + Yi;
      Sum_Y2 := Sum_Y2 + Yi * Yi
    END;
  Correl_Coef :=
      Sqrt(1.0 - SRS /(Sum_Y2 - Sqr(Sum_Y) / Nrow));
  IF Nrow = Ncol THEN
    Nm := 1
  ELSE
    Nm := Nrow - Ncol;
  SEE := Sqrt(SRS / Nm);
  FOR I := 1 TO Ncol DO   { errors in solution }
    Sig[I] := SEE * Sqrt(A[I, I])
END; { Linfit }
```

**Listing 7.5:** Procedure Get_Data and Linfit for the vapor pressure equation (continued).

Notice that the original pressures in procedure Get_Data are given in atmospheres. The procedure then converts these values to the logarithm of the pressure. The X and Y values that are displayed are the temperature and the logarithm of the pressure. To preserve the original pressures, you can use an additional array, which you could call Press. Then you could alter procedure Write_Data so that it displays both the original pressure and the logarithm of the pressure.

In our final example, we will encounter an equation that is unlike the others in this chapter because one of the unknown coefficients is nonlinear—specifically, an exponential. Later, in Chapter 10, we will study algorithms for handling such an equation, but here we will have to be satisfied with a less elegant solution. We will estimate values for the exponent, then find values for the linear coefficients. Then we will change the exponent until we find the best correlation coefficient.

## The Equation
## for the Properties of Superheated Steam

In the previous section, we considered an equation of state for a saturated gas. For that condition, the pressure is a function of temperature. We will now consider an equation of state for a superheated gas. For this example, the temperature is above the saturation temperature or, put another way, the pressure is below saturation. Since

```
I      X        Y      Y CALC     RESID
1    700.0    -20.72   -20.72    -0.00
2    800.0    -16.70   -16.70    -0.01
3    900.0    -13.61   -13.59     0.02
4   1000.0    -11.10   -11.11    -0.00
5   1100.0     -9.08    -9.09    -0.00
6   1200.0     -7.41    -7.41     0.00
7   1300.0     -5.99    -5.99    -0.01
8   1400.0     -4.79    -4.78     0.00
9   1500.0     -3.74    -3.74     0.00
10  1600.0     -2.83    -2.83     0.00

coefficients      errors
 1.81712E+001    5.93731E-001  Constant term
-2.3174E+004     7.73677E+001
-8.8384E-001     7.44396E-002

Correlation coefficient is  1.00000
Press any key for plot
```

**Figure 7.8:** Output from the vapor pressure of liquid lead calculation.

temperature and pressure are independent variables under this condition, we can express the volume as a function of both the temperature and the pressure. Before we get to the program for this calculation, let us consider the three-variable equation of state, which includes a nonlinear coefficient.

## A Three-Variable Equation with a Nonlinear Coefficient

For an ideal gas, the equation of state is

$$PV = RT$$

where V is the molar volume, T is the temperature, P is the pressure, and R is the gas constant. As the pressure increases, the behavior of the gas becomes less and less ideal. Then we must use another equation of state. A common nonideal equation of state is

$$PV = A + BP + CP^2 + DP^3 + \ldots$$

Notice that this equation is just a power series expansion in pressure.

Be careful not to confuse the gas constant R with the vector of residuals r, which we used earlier in the chapter. In our next program, the symbol R is used for the gas constant, and the symbol Resid is used for the residuals.

Since all gases become ideal as the pressure is reduced, an equation of state should merge smoothly with the ideal-gas equation as the pressure approaches zero. For this reason, the value of A in the above equation must be equal to RT. Because fewer polynomial terms are needed at lower pressures, the equation of state can often be written as

$$PV = RT + BP + CP^2$$

The determination of the coefficients B and C is straightforward when the temperature is constant because they are linear. However, if temperature is also a variable, then the coefficients B and C need to be functions of temperature.

Several different equations of state are in common use. All of them, however, are empirical. As an example of a three-variable equation, consider the expression

$$PV = RT + \frac{BP}{T^n} + CP^2$$

In this example, coefficient C is not a function of T, but the original coefficient B becomes the function

$$\frac{B}{T^n}$$

This equation has a nonlinear coefficient, n, so we cannot obtain a solution by the methods discussed in this chapter. On the other hand, if we make an estimate for the coefficient n, then we can determine the remaining linear coefficients. We will begin with an estimate of unity for coefficient n and determine the other coefficients. We can then observe how well the resulting equation represents the original data. Then we will change the value of n and see whether the new equation is better or worse.

## An Equation of State for Steam

We can use the program given in Listing 7.6 to find the coefficients B and C for the published properties of steam. Since the coefficient of the first term on the right is unity, the equation has been rearranged to give

$$PV - RT = \frac{BP}{T^n} + CP^2$$

The data are defined in procedure Get_Data. The temperature is given in degrees Fahrenheit, the pressure in pounds per square inch, and the specific volume in cubic feet per pound mass (these are common engineering units). The temperatures are converted to the absolute Rankine scale by the addition of 460, and pressures are left in pounds per square inch. The gas constant R, which has a value of 85.76 for steam, is multiplied by 144 square inches per square foot.

The data matrix is established in procedure Linfit, as usual. The first column of the matrix contains the pressure divided by the *nth* power of the temperature. Column 2 contains the square of the pressure. The Y vector has the value $PV - RT$.

Notice that this is the first time that we did not put the value of unity in the first column of the matrix. We could, of course, divide the equation by the pressure. The right-hand side would then look like a first-order, straight-line fit. But then the left side would approach infinity when the pressure became very small.

```
PROGRAM Least6;
{ Turbo Pascal program to perform a
  linear least-squares fit
  on the properties of steam
  with Gauss-Jordan elimination routine.
  Separate procedures needed:
        Gaussj and Square }

CONST
  Maxr = 20; { data points }
  Maxc = 4;  { polynomial terms }

TYPE
  Ary   = ARRAY[1..Maxr] OF Real;
  Arys  = ARRAY[1..Maxc] OF Real;
  Ary2  = ARRAY[1..Maxr, 1..Maxc] OF Real;
  Ary2s = ARRAY[1..Maxc, 1..Maxc] OF Real;

VAR
  P, T, V, Y, Y_Calc, Resid: Ary;
    Coef, Sig: Arys;
   Nrow, Ncol: Integer;
  Correl_Coef: Real;

{$I SQUARE.PAS } {Listing 3.2}
{$I GAUSSJ.PAS } {Listing 4.4}

PROCEDURE Get_Data
        (VAR P, T: Ary; { independent variables }
         VAR V   : Ary; { dependent variable }
         VAR Nrow: Integer); { length of vectors }
VAR
  I: Integer;

BEGIN
  Nrow := 12;
  T[1]:=  400; P[1]:=  120; V[1]:=  4.079;
  T[2]:=  450; P[2]:=  120; V[2]:=  4.36;
  T[3]:=  500; P[3]:=  120; V[3]:=  4.633;
  T[4]:=  400; P[4]:=  140; V[4]:=  3.466;
  T[5]:=  450; P[5]:=  140; V[5]:=  3.713;
  T[6]:=  500; P[6]:=  140; V[6]:=  3.952;
  T[7]:=  400; P[7]:=  160; V[7]:=  3.007;
  T[8]:=  450; P[8]:=  160; V[8]:=  3.228;
  T[9]:=  500; P[9]:=  160; V[9]:=  3.44;
  T[10]:= 400; P[10]:= 180; V[10]:= 2.648;
  T[11]:= 450; P[11]:= 180; V[11]:= 2.85;
  T[12]:= 500; P[12]:= 180; V[12]:= 3.042;
  FOR I := 1 TO Nrow DO
    T[I] := T[I] + 460.0  { convert to Rankine }
END; { procedure Get_Data }

PROCEDURE Write_Data;
{ print out the answers }
```

**Listing 7.6:** An equation of state for steam.

```
VAR
  I: Integer;

BEGIN
  WriteLn;
  WriteLn('  I    P      T      V       ',
          'Y     Y CALC    %RES');
  FOR I := 1 TO Nrow DO
    WriteLn(I:3, P[I]:7:1, T[I]:7:1, V[I]:7:3,
      Y[I]:9:2,Y_Calc[I]:9:2,(100.0* Resid[I]/Y[I]):9:2);
  WriteLn;
  WriteLn('coefficients     errors');
  WriteLn(Coef[1]:10,'     ', Sig[1]:10,'  Constant term');
  FOR I := 2 TO Ncol DO
    WriteLn(Coef[I]:10,'     ', Sig[I]:10); { other terms }
  WriteLn;
  WriteLn
    (' Correlation coefficient is ', Correl_Coef:8:5)
END; { Write_Data }

PROCEDURE Linfit
        (P, T, V : Ary;  { independent variables }
         VAR Y      : Ary;  { dependent variable }
         VAR Y_Calc: Ary;  { calculated dependent variable }
         VAR Resid : Ary;  { array of residuals }
         VAR Coef  : Arys; { coefficients }
         VAR Sig   : Arys; { errors in coefficients }
             Nrow  : Integer;  { length of Ary }
         VAR Ncol  : Integer); { number of terms }

{ fit an equation of state through
  Nrow sets of P, T, and V sets of points. }
{ Separate procedures needed:
  Square and Gaussj }

CONST
  R = 85.76; { gas constant for steam }

VAR
    Xmatr: Ary2;  { data matrix }
        A: Ary2s; { coefficient matrix }
        G: Arys;  { constant vector }
    Error: Boolean;
I, J, Nm: Integer;
Power, Yi, Yc, SRS,
SEE, Sum_Y, Sum_Y2: Real;

BEGIN           { procedure Linfit }
  Ncol := 2; { number of terms }
  FOR I := 1 TO Nrow DO
    BEGIN   { Setup X matrix }
      Power := T[I];
      Xmatr[I, 1] := P[I] / Power; { first column }
      Xmatr[I, 2] := Sqrt(P[I]);    { second column }
      Y[I] := V[I] * P[I] - R * T[I] / 144.0
    END;
```

**Listing 7.6:** An equation of state for steam (continued).

```
        Square(Xmatr, Y, A, G, Nrow, Ncol);
        Gaussj(A, G, Coef, Ncol, Error);
        Sum_Y := 0.0;
        Sum_Y2 := 0.0;
        SRS := 0.0;
        FOR I := 1 TO Nrow DO
          BEGIN
            Yi := Y[I];
            Yc := 0.0;
            FOR J := 1 TO Ncol DO
              Yc := Yc + Coef[J] * Xmatr[I, J];
            Y_Calc[I] := Yc;
            Resid[I] := Yc - Yi;
            SRS := SRS + Sqr(Resid[I]);
            Sum_Y := Sum_Y + Yi;
            Sum_Y2 := Sum_Y2 + Yi * Yi
          END;
        Correl_Coef :=
             Sqrt(1.0 - SRS /(Sum_Y2 - Sqr(Sum_Y) / Nrow));
        IF Nrow = Ncol THEN
          Nm := 1
        ELSE
          Nm := Nrow - Ncol;
        SEE := Sqrt(SRS / Nm);
        FOR I := 1 TO Ncol DO    { errors in solution }
          Sig[I] := SEE * Sqrt(A[I, I])
      END; { Linfit }

BEGIN        { main program }
  Get_Data(P, T, V, Nrow);
  Linfit(P,T,V,Y,Y_Calc,Resid,Coef,Sig,Nrow,Ncol);
  Write_Data
END.
```

**Listing 7.6:** An equation of state for steam (continued).

## *Comparing Runs of the Steam Properties Program*

Create a new file with the name LEAST6.PAS, and type the program given in Listing 7.6. Run the program, and check that the results look like Figure 7.9.

It can be seen from Figure 7.9 that the results for a unity value of exponent n are not very good. The correlation coefficient is 92 percent, but some of the calculated values are more than 10 percent from the original data. Let us try another value. Alter the variable named Power near the beginning of procedure Linfit so that the exponent n is 2; that is, make Power equal to the square of the temperature.

Power := Sqr(T[I]);

Rerun the program, and see that the output looks like Figure 7.10.

The resulting fit is better this time. The correlation coefficient is 98 percent, and the calculated points are within 5 percent of the data.

Let us continue to see if we can improve the fit by increasing the exponent to 3. Change the definition of Power to read

Power := T[I] * Sqr(T[I]);

Run the program a third time, and compare the output to Figure 7.11.

The result is definitely better. The fitted values are within 1 percent of the original points, and the correlation coefficient is 99.96 percent. It looks as though we should accept these results. But just to be sure, change the exponent to 4 with the statement

Power := Sqr(Sqr(T[I]));

Run the program again, and compare the results to Figure 7.12. You can see that we have definitely gone too far. The calculated points are

```
I     P      T      U       Y      Y CALC    %RES
1   120.0  860.0  4.079   -22.70   -19.44   -14.36
2   120.0  910.0  4.360   -18.76   -17.43    -7.07
3   120.0  960.0  4.633   -15.77   -15.63    -0.92
4   140.0  860.0  3.466   -26.94   -24.16   -10.30
5   140.0  910.0  3.713   -22.14   -21.82    -1.44
6   140.0  960.0  3.952   -18.45   -19.72     6.85
7   160.0  860.0  3.007   -31.06   -28.98    -6.69
8   160.0  910.0  3.228   -25.48   -26.30     3.24
9   160.0  960.0  3.440   -21.33   -23.90    12.03
10  180.0  860.0  2.648   -35.54   -33.88    -4.67
11  180.0  910.0  2.850   -28.96   -30.86     6.59
12  180.0  960.0  3.042   -24.17   -28.16    16.50

coefficients      errors
-2.62E+002      4.759E+001  Constant term
1.565E+000      6.495E-001

Correlation coefficient is  0.91839
```

**Figure 7.9:** Output from the properties of superheated steam program using a value of 1 for exponent n.

not as close as they were for the previous fit, and the correlation coefficient is farther from unity. Furthermore, the squares of the two standard errors are more than 20 orders of magnitude apart, indicating that the equation is ill-conditioned.

```
 I    P      T      U      Y     Y CALC   %RES
 1  120.0  860.0  4.079  -22.70  -21.49  -5.32
 2  120.0  910.0  4.360  -18.76  -18.03  -3.85
 3  120.0  960.0  4.633  -15.77  -15.10  -4.25
 4  140.0  860.0  3.466  -26.94  -26.01  -3.44
 5  140.0  910.0  3.713  -22.14  -21.98  -0.71
 6  140.0  960.0  3.952  -18.45  -18.56   0.58
 7  160.0  860.0  3.007  -31.06  -30.59  -1.49
 8  160.0  910.0  3.228  -25.48  -25.98   2.00
 9  160.0  960.0  3.440  -21.33  -22.08   3.48
10  180.0  860.0  2.648  -35.54  -35.22  -0.88
11  180.0  910.0  2.850  -28.96  -30.04   3.74
12  180.0  960.0  3.042  -24.17  -25.64   6.08

coefficients      errors
-1.99E+005     1.194E+004  Constant term
9.909E-001     1.803E-001

Correlation coefficient is  0.98901
```

**Figure 7.10:** Output from the properties of superheated steam program using a value of 2 for exponent n.

```
 I    P      T      U      Y     Y CALC   %RES
 1  120.0  860.0  4.079  -22.70  -22.88   0.78
 2  120.0  910.0  4.360  -18.76  -18.80   0.22
 3  120.0  960.0  4.633  -15.77  -15.52  -1.59
 4  140.0  860.0  3.466  -26.94  -26.97   0.13
 5  140.0  910.0  3.713  -22.14  -22.21   0.35
 6  140.0  960.0  3.952  -18.45  -18.39  -0.32
 7  160.0  860.0  3.007  -31.06  -31.09   0.10
 8  160.0  910.0  3.228  -25.48  -25.65   0.68
 9  160.0  960.0  3.440  -21.33  -21.28  -0.23
10  180.0  860.0  2.648  -35.54  -35.22  -0.90
11  180.0  910.0  2.850  -28.96  -29.10   0.49
12  180.0  960.0  3.042  -24.17  -24.19   0.06

coefficients      errors
-1.39E+008     1.502E+006  Constant term
2.999E-001     2.520E-002

Correlation coefficient is  0.99963
```

**Figure 7.11:** Output from the properties of superheated steam program using a value of 3 for exponent n.

We could try to improve the fit by choosing noninteger exponents close to the value of 3 or by making coefficient C a function of temperature. However, at this point, we will be satisfied with the reasonably good fit provided with an exponent of 3.

## Summary

This chapter has illustrated the idea of writing a general curve-fitting program. We progressed through several versions including:

- A straight-line fit

- A parabolic curve fit

- A more direct parabolic curve fit, using the matrix approach

- A general polynomial curve fit, in which the polynomial order can be adjusted

- Curves fitting nonpolynomial equations

Remember, the restriction on all these versions is that the coefficients must be linear. Although we found a way around this restriction in our last example, we must still develop a general method for dealing with nonlinear coefficients. We will study nonlinear curve-fitting equations in Chapter 10.

```
    I    P      T      U      Y      Y CALC    %RES
    1  120.0  860.0  4.079  -22.70  -23.69    4.38
    2  120.0  910.0  4.360  -18.76  -19.32    3.02
    3  120.0  960.0  4.633  -15.77  -16.00    1.45
    4  140.0  860.0  3.466  -26.94  -27.46    1.94
    5  140.0  910.0  3.713  -22.14  -22.36    1.02
    6  140.0  960.0  3.952  -18.45  -18.49    0.19
    7  160.0  860.0  3.007  -31.06  -31.22    0.51
    8  160.0  910.0  3.228  -25.48  -25.39   -0.34
    9  160.0  960.0  3.440  -21.33  -20.96   -1.74
   10  180.0  860.0  2.648  -35.54  -34.96   -1.62
   11  180.0  910.0  2.850  -28.96  -28.41   -1.89
   12  180.0  960.0  3.042  -24.17  -23.43   -3.08

 coefficients      errors
 -9.85E+010      3.648E+009   Constant term
 -1.91E-001      6.828E-002

 Correlation coefficient is  0.99571
```

**Figure 7.12:** Output from the properties of superheated steam program using a value of 4 for exponent n.

# *Exercises*

**7-1.** If you know the thermodynamic activity of one component of a binary solution, you can calculate the activity of the other component. The logarithm of the activity coefficient $\gamma$ (the ratio of activity to mole fraction), is fitted to a power series in mole fraction of the other component. Notice that the series begins with the second-order term.

$$\ln \gamma_1 = AX_2^2 + BX_2^3 + CX_2^4 + \ldots$$

The activity, $A_{Ni}$, of nickel in iron-nickel solutions at 1500 kelvins as a function of mole fraction of nickel, $X_{Ni}$, has been reported as the following:

| $X_{Ni}$ | 1 | .9 | .8 | .7 | .6 | .5 | .4 | .3 | .2 | .1 |
|---|---|---|---|---|---|---|---|---|---|---|
| $A_{Ni}$ | 1 | .89 | .766 | .62 | .485 | .374 | .283 | .207 | .136 | .067 |

Make a copy of the program given in Listing 7.2. Alter procedure Get_Data so that the number of data points, Nrow, is changed to 10. Change the FOR loop so that X is defined as the mole fraction of iron:

```
X[I] = 0.1 * (i - 1)
```

Change Y so that it contains the corresponding nickel activities. Alter procedure Linfit so that the power series will begin with the second-order term:

```
Xmatr[I,1] := Sqr(Xi)
Xmatr[i,2] := Sqr(Xi) * Xi
```

The dependent variable can be converted within this same loop:

```
Y[I] := Ln(Y[I]) / (1.0 - X[I])
```

In this step, the activity coefficient is obtained from a ratio of the activity to the mole fraction. The dependent variable is the logarithm of the ratio. Determine the coefficients A, B, and C.

**7-2.** The vapor pressure of mercury is reported to be:

| Temp (°C) | Pressure (torr) |
|---|---|
| 0 | 0.000185 |
| 50 | 0.01267 |
| 100 | 0.2729 |
| 150 | 2.2807 |
| 200 | 17.287 |
| 250 | 74.375 |
| 300 | 246.80 |
| 350 | 672.69 |
| 400 | 1574.1 |

Find the coefficients A, B, and C for a least-squares fit to the equation

$$\ln P = A + B/T + C \ln T$$

The X data can be generated with the input loop:

X[I] := (i-1) * 50 + 273.15

and the Y data can be defined directly, then converted to the logarithm in a FOR loop. Use the vapor pressure program developed in Listing 7.5.

**7-3.** Find the coefficients A, B, and C to

$$C_p = A + BT + C/T^2$$

for the heat capacity of graphite. Use the program given in Listing 7.4.

| Temperature (kelvins) | Heat Capacity (cal/deg mole) |
|---|---|
| 300 | 2.08 |
| 400 | 2.85 |

| | |
|---|---|
| 500 | 3.50 |
| 600 | 4.03 |
| 700 | 4.43 |
| 800 | 4.75 |
| 900 | 4.98 |
| 1000 | 5.14 |
| 1100 | 5.27 |
| 1200 | 5.42 |

# Solution of
# Equations by
# Newton's Method

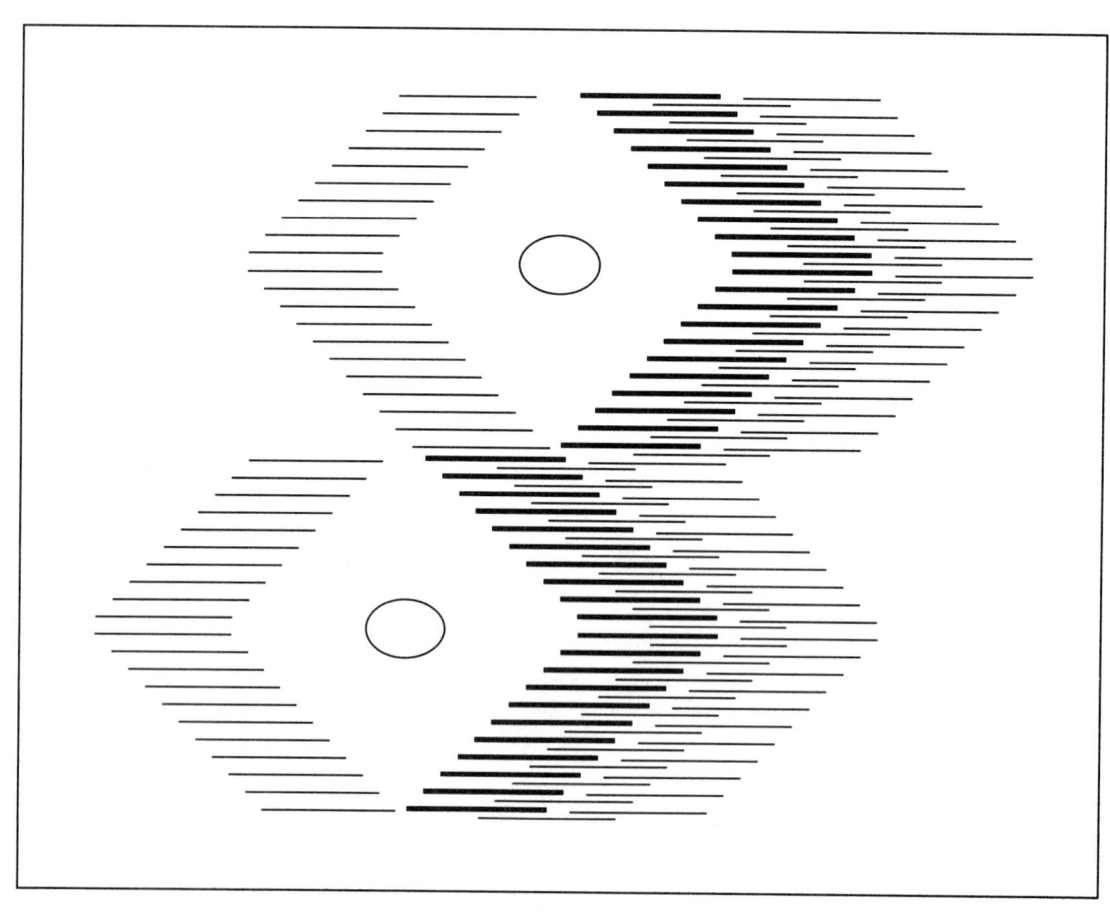

# Chapter 8

In this chapter, we will develop a computer program for solving equations by a technique known as Newton's method, or the Newton-Raphson method. This technique is especially suitable for finding the roots of "well-behaved" functions. (Well-behaved functions change slowly and smoothly and usually have only one solution, at least in the region of interest. Fortunately, we can model most applications with well-behaved functions.) Newton's method has many applications, and we will use it later in the book when we return to the topic of curve fitting with nonlinear coefficients.

First we will outline the mathematical formulation of Newton's method; then we will consider a series of progressively more sophisticated Turbo Pascal versions of the technique. We will study the two pitfalls of the method—the cases where the tangent to the curve has a zero slope and where successive approximations fail to converge on a root—and consider ways of dealing with them. Finally, we will use our program to solve a practical application.

# *Formulating Newton's Method*

Let us begin by considering a general equation of the form

$$f(x) = 0 \qquad\qquad (1)$$

This equation might have one solution, several solutions, or none at all. That is, there may be particular values of x that make the equation equal to zero. These values are called the *roots*, or *solutions*, of the equation. For other values of x, the function will not be zero.

Sometimes we can solve such an equation explicitly. For example, the expression

$$x^2 - 4 = 0$$

can be converted to:

$$x^2 = 4$$

which has the solutions:

$$x = 2 \quad\text{and}\quad x = -2$$

But sometimes an equation cannot be solved so easily. As an example, consider the vapor pressure equation that we discussed in the previous chapter:

$$\ln P = A + \frac{B}{T} + C \ln T$$

We can easily find the vapor pressure of lead at, say, 1000 kelvins by solving the expression

$$\ln P = A + \frac{B}{1000} + C \ln 1000$$

But suppose that we want to find the temperature that corresponds to a lead vapor pressure of 0.1 atmosphere. We want to solve the equation

$$\ln 0.1 = 18.19 - \frac{23180}{T} - 0.8858 \ln T$$

This nonlinear equation cannot be solved explicitly for the temperature. However, we can use Newton's method to approximate the answer to as high a precision as we desire.

For the general case, we write the equation

$$y = f(x)$$

We are interested in the values of x when y is equal to zero; that is, we want to determine the points where the curve of the function crosses the x axis.

Consider, for example, an equation with two solutions:

$$y = f(x) = x^2 - 4$$

This curve crosses the x axis at two places, $+2$ and $-2$, as shown in Figure 8.1.

An example of an equation with one solution is

$$y = f(x) = x^2$$

This curve is tangent to the x axis at the origin, corresponding to a single root, x = 0, as shown in Figure 8.2.

Finally, an equation with no real solution is

$$y = f(x) = x^2 + 4$$

As shown in Figure 8.3, this equation does not cross the x axis at all, and so it has no real roots.

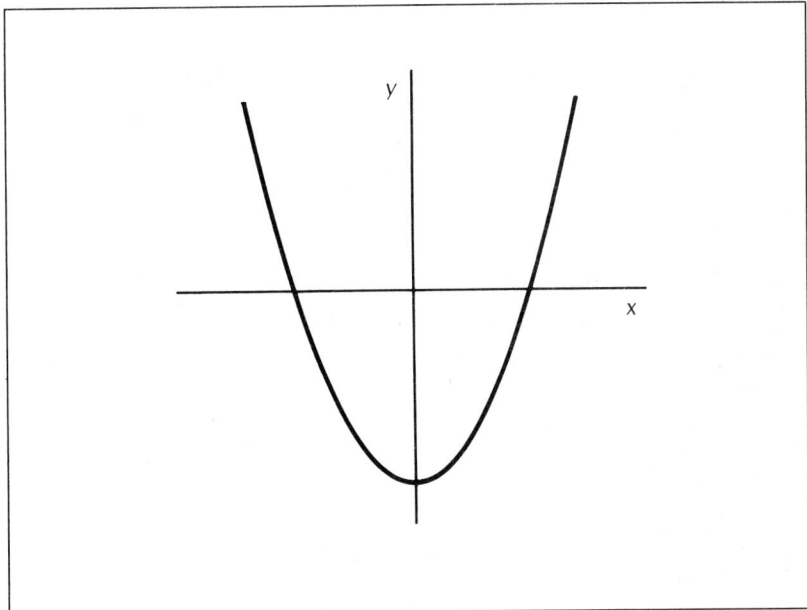

**Figure 8.1:** A function with two solutions.

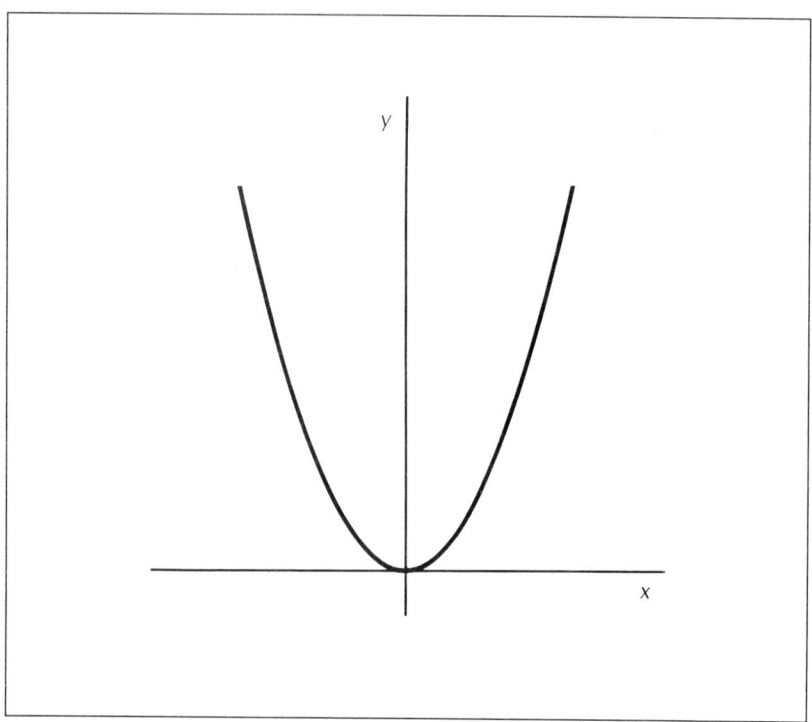

**Figure 8.2:** A function with one solution.

Let us explore the behavior of the general function $y = f(x)$ near a root. We might find that it looks like the curve of Figure 8.4. The function crosses the x axis at a root because the relationship $y = f(x) = 0$ is satisfied there.

## Approximating Values

We start Newton's method with an approximate value for x, say $x_1$, that is near a root. We can determine the corresponding value of y by the equation $y_1 = f(x_1)$. This represents a point on the curve that is not, in general, a root. We now construct a tangent to f(x) at this point on the curve and extend it until it intersects the x axis. The next approximation, $x_2$, is at this intersection on the x axis, as illustrated in Figure 8.5. Notice that, in this example, the second approximation, $x_2$, is closer to the root than the first approximation, $x_1$. Thus, we have refined our original approximation.

We now repeat the process. We evaluate the function at $x = x_2$ to obtain the corresponding value of y, $y_2 = f(x_2)$. The value of $y_2$ is

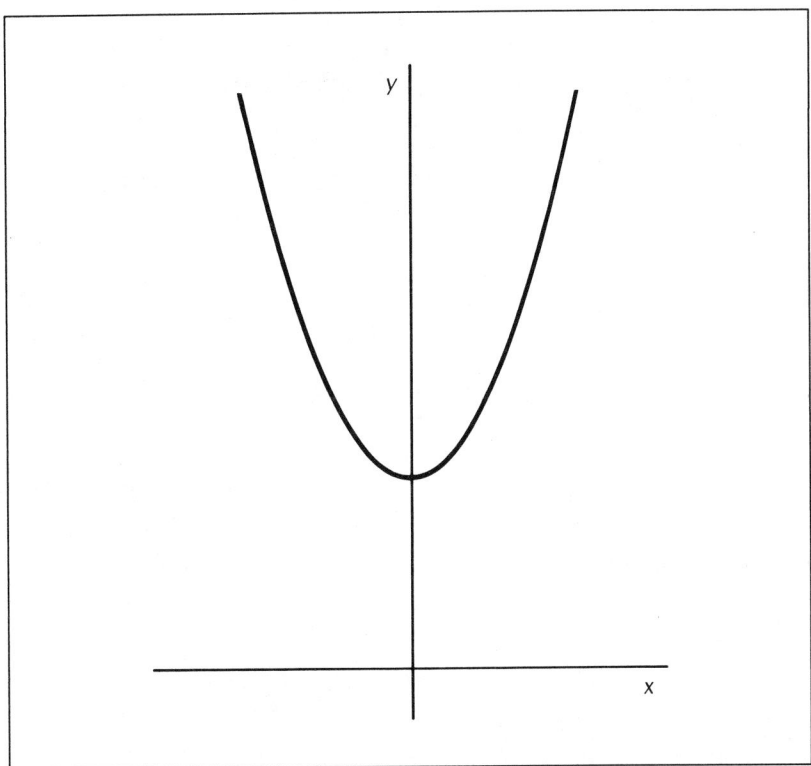

**Figure 8.3:** A function with no real roots.

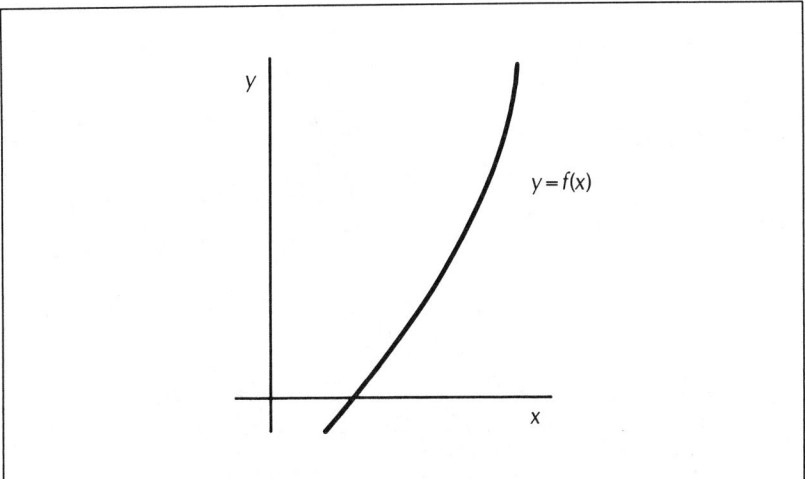

**Figure 8.4:** f(x) = 0 is satisfied where the curve crosses the x axis.

smaller than the value of $y_1$, showing that we are closer to the root. We construct another tangent, this time at the point $(x_2, f(x_2))$. The intersection of the new tangent with the x axis gives the value of $x_3$, the third approximation of x. We continue in this way, improving the value of x until we are as close to the root as we want.

Let us go back and review the first step in more detail. The initial approximation, $x_1$, gives rise to $y_1 = f(x_1)$. The tangent constructed at $y_1$ has a slope of

$$f'(x_1) = \frac{y_1}{x_1 - x_2} \tag{2}$$

Because $y_1 = f(x_1)$, Equation 2 can be expressed as

$$x_2 = x_1 - \frac{f(x_1)}{f'(x_1)} \tag{3}$$

or more generally as

$$x_{i+1} = x_i - \frac{f(x_i)}{f'(x_i)} \tag{4}$$

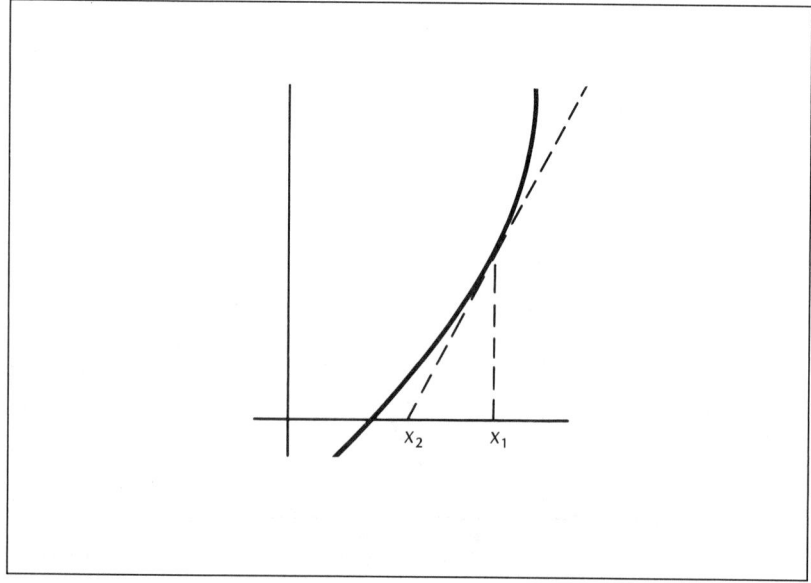

**Figure 8.5:** The tangent crosses the x axis closer to the root than the original
approximation for x.

where $x_i$ is the *ith* approximation. Equation 4 is the usual form of Newton's method. It can be an ideal technique for finding a root for a real-life equation.

Equations that deal with the behavior of real things typically have only one meaningful root, at least in the area of interest. The other roots of such equations will usually be negative, zero, or complex. Furthermore, the approximate value of the answer may be known. For example, the ideal-gas law can provide a first approximation to a more complicated equation of state.

Now that we have arrived at an equation for the general form of Newton's method, developing our program will be relatively easy.

## *Programming Newton's Method*

We begin by applying Newton's method to a simple problem, one for which we already know the answer. The equation we will solve is

$$x^2 = 2$$

or

$$x^2 - 2 = 0 \qquad\qquad (5)$$

The positive solution to this equation is the square root of 2. First, we define the function

$$y = f(x) = x^2 - 2 \qquad\qquad (6)$$

and its derivative

$$\frac{dy}{dx} = f'(x) = 2x \qquad\qquad (7)$$

Our first attempt at a program for Newton's method is shown in Listing 8.1. The algorithm itself is contained in procedure Newton. A separate procedure, called Func, is used for the calculation of the function (Equation 6) and its derivative (Equation 7). The body of the main program provides the first approximation; it calls procedure Newton to find the solution and to display the answer.

As a matter of good programming practice, procedures and functions that are used to calculate values should not display intermediate results. However, in this particular case, it is instructive to observe the successive values of x, f(x), and f'(x) during convergence. For this reason, there is a WriteLn statement in procedure Newton. You may want to remove this statement when the program is working properly.

## *Tolerance*

There are a few additional matters we need to consider. Successive approximations are provided by a REPEAT/UNTIL loop in procedure Newton. This loop continues until two successive values are within the desired tolerance. We are not interested in whether the difference (Dx)

```pascal
PROGRAM Newton1;
{ solve equation by Newton's method }

VAR
    X, X2: Real;
  AllDone: Boolean;
    Error: Boolean;

PROCEDURE Func(X: Real;
      VAR Fx, Dfx: Real);
BEGIN
  Fx := X * X - 2.0;
  Dfx := 2.0 * X
END;  { Func }

PROCEDURE Newton(VAR X: Real);

CONST
  Tol   = 1.0E-6;

VAR
  Fx, Dfx, Dx, X1: Real;

BEGIN   { Newton }
  REPEAT
    X1 := X;
    Func(X, Fx, Dfx);
    Dx := Fx / Dfx;
    X := X1 - Dx;
    WriteLn
      ('X= ',X1:8:5,', Fx= ',Fx:12,', Dfx= ',Dfx:8:5)
  UNTIL  Abs(Dx) <= Abs(Tol * X)
END;   { Newton }

BEGIN        { main program }
  WriteLn;
  X := 2.0;  { first approximation }
  Newton(X);
  WriteLn;
  WriteLn('The solution is ', X:9:5);
  WriteLn
END.
```

**Listing 8.1:** Newton's method, version one.

between two successive approximations has a negative or a positive value. We are only concerned with the magnitude. For this reason, we must be careful to take the absolute value of the comparison.

Furthermore, we are not interested in the actual difference, but only the relative difference. Suppose, for example, that we want our answers to be accurate to one part in a million, that is, 1 part in $10^6$. If a particular solution has a value of unity, then two successive values must be closer than $10^{-6}$. If, however, the solution itself has a value of $10^{-6}$, then, two successive values must differ by no more than $10^{-12}$. We therefore choose a relative criterion rather than an absolute one for terminating the iteration process.

### *Generalizing Procedure Calls*

Another matter we should consider is the relationship of procedure Newton to procedure Func. Procedure Newton gives directions for carrying out the operation described by Equation 4. It is independent of the actual function that it is operating on. For this reason, procedure Func, used to obtain the function and its derivative, should be a separate entity.

Ideally, the procedure name Func, within Newton, should be a replaceable parameter. The actual procedure name should be passed to procedure Newton as a parameter during execution. With this arrangement, procedure Newton could be directed to solve one equation at one point and another equation at another point. This technique is easily accomplished in Fortran. However, Pascal compilers, including Turbo Pascal, do not allow procedure names to be passed as parameters to other procedures.

If two or more different equations are to be solved in the same program, you will have to provide additional, separate copies of procedure Newton, each with a different name. Another, more complicated, approach would be to use a CASE statement in procedure Func. This approach would select one function on the first call, a second function on the next call, and so on.

### *Running the Newton's Method Program*

Type the program shown in the Listing 8.1, give it the name NEWTON1.PAS, and run it. The first approximation to the solution is chosen to be 2. Then, successive approximations should converge to the square root of 2 after several iterations. The results will look like Figure 8.6.

In the following sections, we will refine this program. The first change will allow us to enter different first-approximation values for the root. This facility is especially important for studying equations that have more than one root.

## Adding User Input for the First Approximation

When the first version of Newton's method is working properly, you can begin to add new features. Make a copy of the first version using the name NEWTON2.PAS. Change the main program (the part within the final BEGIN and END block) so that it looks like Listing 8.2.

```
        x=  2.00000,  fx= 2.00000E+000 , dfx=  4.00000
        x=  1.50000,  fx= 2.50000E-001 , dfx=  3.00000
        x=  1.41667,  fx= 6.94444E-003 , dfx=  2.83333
        x=  1.41422,  fx= 6.00730E-006 , dfx=  2.82843
        x=  1.41421,  fx= 4.51061E-012 , dfx=  2.82843

        The solution is   1.41421
```

**Figure 8.6:** Output from the program to calculate the positive root of $f(x) = x^2 - 2$.

```
BEGIN       { main program }
  AllDone := False;
  REPEAT
    WriteLn;
    Write(' First guess: ');
    ReadLn(X);
    IF X < -19.0 THEN
      AllDone := True
    ELSE
      BEGIN
        Newton(X);
        WriteLn;
        WriteLn('The solution is ', X:9:5);
        WriteLn
      END
  UNTIL AllDone
END.
```

**Listing 8.2:** The main program for the second version.

Run the new version. Unlike the first version, which uses the value of 2 as the first guess to Newton's method, this version asks you to enter the first approximation from the keyboard. It then displays the successive approximations, along with the value of the function and its derivative, as the first version did. At the conclusion of the task, the program displays the root. Then it begins again, and you are asked to enter another first approximation.

## *Running the Program to Find the Second Root*

Start with a first approximation of 2. The results will be the same as they were for the first version. Then, for the second cycle, try the value of 1. This first approximation is on the other side of the root, but the program should again find the square root of 2 in a relatively few steps.

Try the value of $-2$ for the third cycle. Notice that the process converges on a different root this time. There are two solutions to the equation

$$x^2 - 2 = 0$$

We found the other root this time by giving a negative first approximation.

Let us investigate what happens when the first approximation is near the midpoint of the two roots. Try a value of 0.0001. In this example, the program requires many steps to produce the answer.

Finally, try a new first guess of 0. The curve $y = f(x)$ has zero slope at this point. As a result, Turbo Pascal terminates the program because there is an attempt to divide by zero. We will correct this problem in the next version.

## *A Test for Zero Slope*

When the derivative, or slope, of a function is zero, the final term in Equation 4 becomes infinite. The slope to the function will never cross the x axis because the two lines are parallel, as shown in Figure 8.7.

We will now add some instructions for testing the slope. One way to do this is to define a small number such as

```
CONST Small = 1.0E - 16;
```

Then, the slope can be tested with the statement

```
IF Abs( Dfx ) < Small THEN . . .
```

One problem with this approach, however, is that the value assigned to Small must be consistent with the particular version of Pascal being used. That is, the value of Small may have to be chosen carefully.

In our program, shown in Listing 8.3, we will use a more straight-forward method to check for zero slope:

**IF** Dfx = 0.0 **THEN** . . .

If the slope becomes zero, the program displays an error message and sets an error flag. Otherwise, the process continues normally.

Copy the previous program, giving it the name NEWTON3.PAS. Change it to look like Listing 8.3. The new main program tests the error flag after each call to procedure Newton. If the flag is not set, then there is no error, and the solution is displayed.

## Running the Program with the Slope Test

Run the new version. Enter the initial values of 2, 1, and −1, as before, to see that the program behaves properly. Then, give an initial

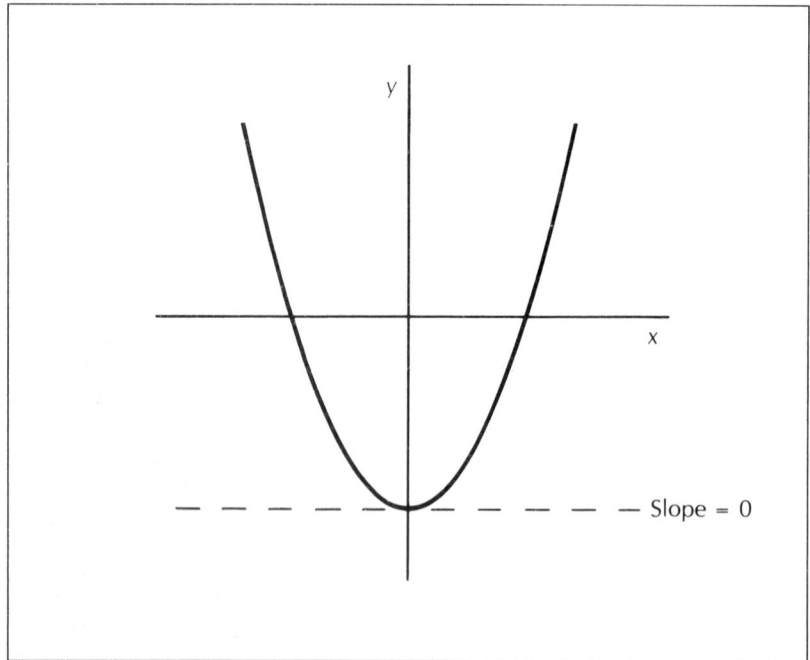

**Figure 8.7:** At f′(x) = 0 the tangent is parallel to the x axis.

```
PROGRAM Newton3;
VAR
  X, X2  : Real;
  AllDone: Boolean;
  Error  : Boolean;

PROCEDURE Func (X: Real;
      VAR Fx, Dfx: Real);

BEGIN
  Fx := X * X - 2.0;
  Dfx := 2.0 * X
END; { Func }

PROCEDURE Newton(VAR X: Real);

CONST
  Tol   = 1.0E-6;

VAR
  Fx, Dfx, Dx, X1: Real;

BEGIN  { Newton }
  Error := False;
  REPEAT
    X1 := X;
    Func(X, Fx, Dfx);
    IF Dfx = 0.0 THEN
      BEGIN
        Error := True;
        WriteLn('ERROR: slope zero')
      END
    ELSE
      BEGIN
        Dx := Fx / Dfx;
        X := X1 - Dx;
        WriteLn
          ('X= ',X1:8:5,', Fx= ',Fx:12,' , Dfx= ',Dfx:8:5)
      END
  UNTIL Error OR (Abs(Dx) <= Abs(Tol * X))
END; { Newton }
BEGIN     { main program }
  AllDone := False;
  REPEAT
    WriteLn;
    Write (' First guess: ');
    ReadLn( X);
    IF X < -19.0 THEN AllDone := True
    ELSE
      BEGIN
        Newton(X);
        WriteLn;
        IF NOT Error THEN
          WriteLn('The solution is ', X:9:5);
        WriteLn
      END
  UNTIL AllDone
END.
```

**Listing 8.3:** Newton's method with a test for zero slope.

value of 0. Instead of Turbo Pascal terminating the program and displaying an error message, as with the previous version, the program now displays our error message and requests another first approximation. You can abort the program by entering a first approximation that is less than $-19$ or by pressing Ctrl-C.

Our next task is to make sure that the program will display an error message and terminate after an appropriate number of iterations if the approximations do not converge on a root.

### *Failure to Converge*

Sometimes, Newton's method will not converge on a root after a reasonable number of iterations. One possibility is that successive approximations are oscillating around a complex root, as illustrated in Figure 8.8. The first approximation, $x_1$, gives rise to the second value, $x_2$. But $x_2$ then produces the original value of $x_1$ for the third approximation. In this example, the process will never terminate.

Another possibility is that an approximation is very far from a root. This can occur even though the first guess appears to be close to a root. We can observe this behavior with our previous version of the

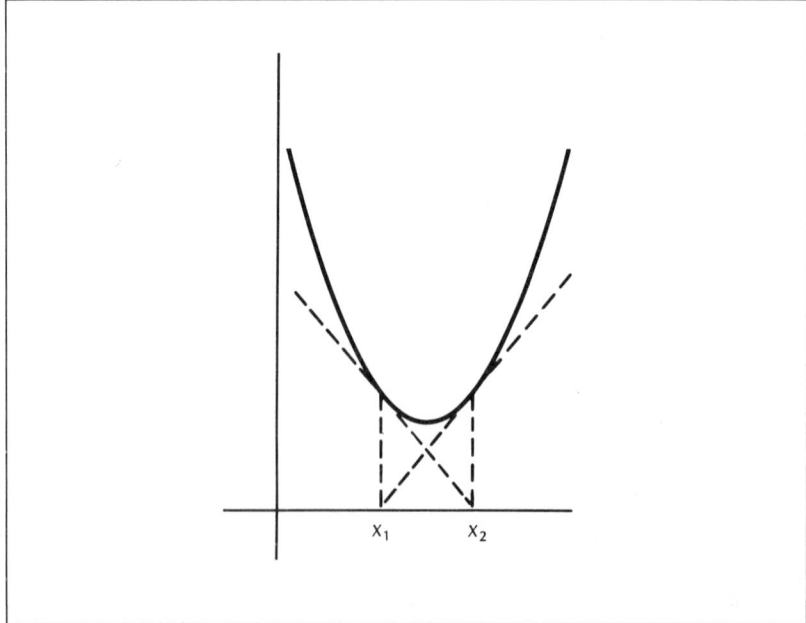

**Figure 8.8:** A complex root.

Newton's method if we run the program again and give a first approximation of 0.00001. This produces a second value of 20,000, which is quite an overshot. Each successive approximation will then be about one-half of the previous value. The program eventually obtains a solution; however, more than 20 iterations are required for convergence.

We can guard against failure to converge in either of these cases by adding a loop counter to procedure Newton. We can then terminate the procedure if convergence does not occur after, say, 20 iterations.

Make a copy of your program, giving it the name NEWTON4.PAS. Alter procedure Newton so that it looks like the version in Listing 8.4.

```
PROCEDURE Newton(VAR X: Real);

CONST
   Tol  = 1.0E-6;
   Max = 20;

VAR
   Fx, Dfx, Dx, X1: Real;
   I: Integer;

BEGIN  { Newton }
   Error := False;
   I := 0;
   REPEAT
     I := I + 1;
     X1 := X;
     Func(X, Fx, Dfx);
     IF Dfx = 0.0 THEN
       BEGIN
         Error := True;
         X := 1.0;
         WriteLn('ERROR: slope zero')
       END
     ELSE
       BEGIN
         Dx := Fx / Dfx;
         X := X1 - Dx;
         WriteLn
           ('X= ',X1, ', Fx= ',Fx, ', Dfx= ',Dfx)
       END
   UNTIL
     Error  OR
     (I >= Max) OR
     (Abs(Dx) <= Abs(Tol * X));
   IF I >= Max THEN
     BEGIN
       WriteLn
         ('ERROR: no convergence in ',
          Max, ' loops');
       Error := True
     END
END; { Newton }
```

**Listing 8.4:** Newton's method with a loop counter.

Run the new version, giving an initial approximation of 500000. This first guess is so far from the root that it will take many cycles to converge. The program stops the iteration process after 20 loops and displays an error message. Notice that the solution at this point is 1.42174, close to the correct value. Next, try a first approximation of 2 and check that the square root of 2 is displayed.

We have developed and refined our program using a simple, predictable function. Now we are ready to use it to find the roots of some more difficult functions.

## *Programming the Solutions to Other Equations*

Consider the nonlinear equation

$$e^x = 4x$$

This equation, unlike the one we solved previously in this chapter, cannot be solved explicitly for x, so it is a suitable candidate for our Newton's method program. The corresponding function and its derivative are

$$f(x) = e^x - 4x$$
$$f'(x) = e^x - 4$$

Copy the Newton program, giving it the name NEWTON5.PAS. Change procedure Func so that it looks like the version in Listing 8.5. The

```
PROCEDURE Func(X: Real;
      VAR Fx, Dfx: Real);

VAR
   E: Real;

BEGIN
   E := Exp(X);
   Fx := E - 4.0 * X;
   Dfx := E - 4.0
END; { Func }
```

**Listing 8.5:** Procedure Func for $e^x = 4x$.

variable E is introduced into this procedure so that the exponent of X will not have to be calculated twice. Recall that

$$\frac{de^x}{dx} = e^x$$

Run the new version, and enter a first approximation of 4. The program converges on the value 2.153. This is one of the two roots. You can find the other root by giving a first approximation of 0. The result this time is a value of 0.357.

## A Function with Many Roots

Let us explore the several roots of the equation:

$$Sin(x) = \frac{x}{10}$$

Copy the previous program and give it the name NEWTON6.PAS. Change procedure Func to the version shown in Listing 8.6.

Run this version, giving a first estimate of 1.0. This will converge on the root at 0. Then, try a first approximation of $-1.0$. This too should converge on the root at 0. There are several other roots to this equation. The table in Figure 8.9 gives first approximations and the corresponding roots. Try each one to verify that your Newton's method program is working properly.

```
PROCEDURE Func(X: Real;
          VAR Fx, Dfx: Real);

BEGIN
   Fx   := Sin(X) - 0.1 * X;
   Dfx  := Cos(X) - 0.1
END;  { Func }
```

**Listing 8.6:** The solution of $Sin(X) = x/10$.

| First Approximation | Root |
|:---:|:---:|
| − 1 | 0 |
| 1 | 0 |
| 4 | 2.852.. |
| 4.3 | 7.068.. |
| 4.5 | 0 |
| 4.7 | − 8.423.. |
| 5 | − 2.852.. |
| 6 | 7.068.. |
| 9 | 8.423.. |

**Figure 8.9:** The roots of Sin(x) − x/10 = 0.

## Solving the Vapor Pressure Equation

We are now ready to solve the vapor pressure equation. We will write our function and its derivative as:

F(t)  = A + B/T + C Ln T − Ln P
Df(t) = − B/T2 + C/T

Remember that A, B, C, and P are constants. Make a copy of the previous program, giving it the name NEWTON7.PAS. Alter procedure Func so that it looks like the version in Listing 8.7. Notice that the letter T is used in place of X. We can make this change because the parameters to procedure Func are dummy variables.

With this version of Func, we can find the temperature that corresponds to a vapor pressure of 0.01 atmosphere. The logarithm of this pressure, − 4.60517, is entered directly as a constant.

Run the new version, entering a first approximation of 500 kelvins. The program will converge to a temperature of 1416 kelvins in seven steps.

## Summary

Using a familiar function at first, we saw how easy it was to write and then improve upon a Turbo Pascal version of Newton's method for finding the roots of an equation. We used our program to solve some

```
        PROCEDURE Func(T: Real;
             VAR Ft, Dft: Real);
          { the vapor pressure of lead }

        CONST
          A = 18.19;
          B = -23180.0;
          C = -0.8858;
          LogP = -4.60517; { Ln(.01) }

        BEGIN
          Ft := A + B / T + C * Ln(T) - LogP;
          Dft := - B /(T * T) + C / T
        END;  { Func }
```

**Listing 8.7:** Solution of the vapor pressure equation.

more complex exponential and trigonometric functions. Finally, we solved an expression that is used to describe vapor pressure.

## *Exercises*

**8-1.** The degree of dissociation, X, of hydrogen sulfide gas can be described by the equation

$$(1 - PK^2)X^3 - 3X + 2 = 0$$

where K is the equilibrium constant and P is the total pressure in atmospheres. For a temperature of 2000 kelvins, K = 0.608. Use Newton's method to find the degree of dissociation when the pressure is 1 atmosphere. Since X only has meaning over the range 0 through 1, begin with an approximation of 0.5. Explore the consequences of initial guesses of 0, 1.0, and 1.2.

**8-2.** van der Waals equation of state is

$$(P + a/v^2)(v - b) = RT$$

where P is the pressure, v is the molar volume, T is the temperature, and R is the gas constant. Coefficient a is a measure of the bonding force, and coefficient b is a volume correction. When the pressure is in atmospheres, the volume in liters, and the temperature in kelvins,

R has a value of 0.082 liter-atm/kelvins mole. Coefficient a is in units of pressure times molar volume squared, and b is in units of molar volume. The coefficients a and b have been extensively tabulated for common gases. For example, the values for toluene are

and
$$a = 24.06 \ 1^2 \ \text{atm/mole}$$
$$b = 0.1463 \ 1/\text{mole}$$

Use Newton's method to solve the van der Waals equation. Find the molar volume of toluene at the boiling point of 110° Celsius (383 kelvins) and a pressure of 1 atmosphere.

Encode the van der Waals equation into procedure Func; use the form shown in Listing 8.7. Alter the main program so that the first approximation to the volume, X, is obtained from the ideal gas law

$$V = RT/P$$

Run the program to find the van der Waals volume.

**8-3.** Use Newton's method to solve the van der Waals equation as with the previous problem, but have the program read the temperature and pressure from the keyboard. Display both the ideal gas volume and the van der Waals volume.

# Numerical
# Integration

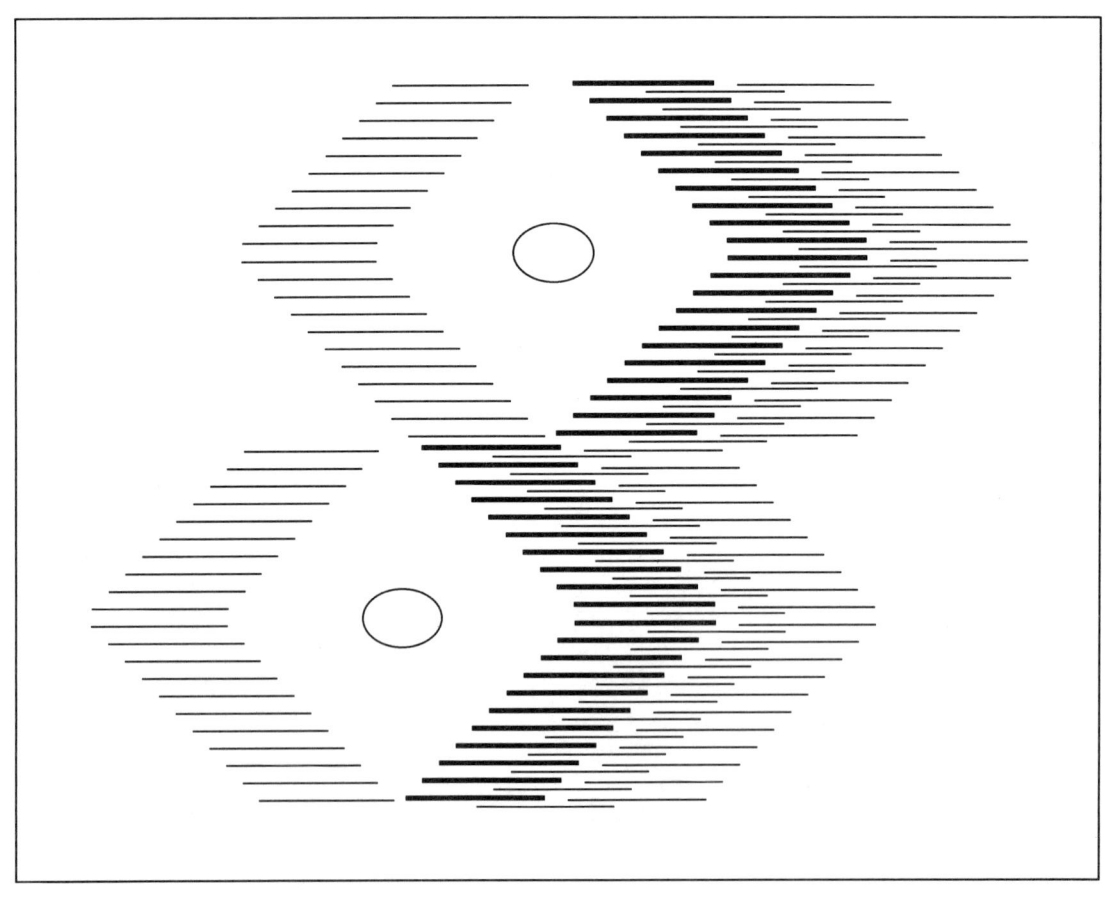

*Chapter* **9**

When a mathematical expression cannot be directly integrated with calculus, then we must obtain its integral by numerical methods. In this chapter, we will develop three different techniques for numerical integration: the trapezoidal, Simpson's, and Romberg methods. These methods directly determine the area beneath the curve between two particular values of the independent variable. They use progressively smaller "panels" of measurable area to divide the area beneath the curve and then sum the panels to approximate the total area.

We will develop Turbo Pascal programs for each method and compare the accuracy of their output. One of the functions that we will solve using both the Simpson's and Romberg methods is related to the normal distribution function, which we will be returning to in Chapter 11. Finally, we will explore a method for integrating a function that approaches infinity at one of its limits.

Let us begin with a brief description of the definite integral and the methods of evaluating it.

# The Definite Integral

The evaluation of the *definite integral*

$$\int_a^b f(x)dx = F(b) - F(a)$$

where $F'(x) = dF(x)/dx = f(x)$ can be interpreted as the area under the curve of the function f(x) from the limit a to the limit b, as illustrated in Figure 9.1. The integration can be straightforward for certain functions, but very difficult for others. For example, the power series

$$1 + 2x + 3x^2$$

can be integrated term by term, and then evaluated between the limits of a and b to give

$$b + b^2 + b^3 - a - a^2 - a^3$$

There are various methods for integrating more complicated functions. Some can be integrated by reference to integration tables found in mathematics, some can be *integrated by parts*, and others can be integrated by replacing the original function with an infinite series, or an *asymptotic expansion*. In cases where the integrand is not a proper function at all, but simply a collection of experimental data, it may be possible to fit the experimental data to a function that can be integrated. Nevertheless,

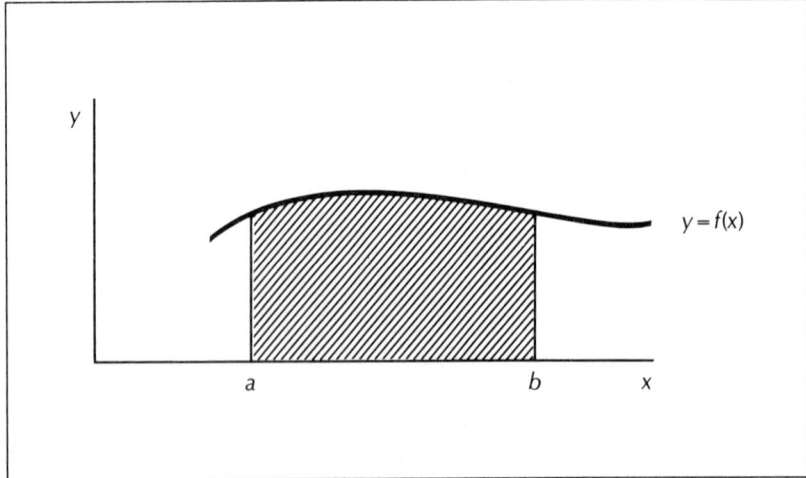

**Figure 9.1:** The area under the curve y = f(x).

there will be times when it will be very difficult (or impossible) to evaluate the integrand analytically. However, if you know the limits, then the approximation method of *numerical integration* may provide an acceptable solution.

The methods commonly used for numerical integration typically involve the substitution of an easily integrated function for the original function. The new function may be a polynomial, such as a straight line or parabola, or it may consist of transcendental functions, such as sines and cosines. The accuracy of the resulting calculation depends on how well the substituted function approximates the original function.

In the next section, we will examine a general method of function substitution and review a specific, simple example.

## *The Trapezoidal Method*

One of the simplest techniques of numerical integration is the *trapezoidal method*, which approximates the original function by a set of straight lines. The region to be integrated is divided into uniformly spaced panels. The panel width, $\Delta x$, is

$$\Delta x = \frac{b - a}{n}$$

where n is the number of panels, and a and b are the integration limits. If the entire integral is fitted with a single straight line, that is, if there is only one panel (as illustrated in Figure 9.2), then the calculated area is

$$\frac{(b - a)[f(a) + f(b)]}{2}$$

In this formula, f(a) is the value of the function at the left limit, and f(b) is the value at the right limit.

For the more general case, the area is divided into $n$ panels. An interior panel is bounded on the left by a vertical line at $x_i$, and on the right by a vertical line at $x_i + \Delta x$. The lower edge of the panel is marked by the x axis. The original curve at the top of the panel is replaced by a straight line that generally will not have a zero slope (that is, not parallel to the x axis). The resulting panel has the shape of a trapezoid.

The panels can be numbered from 1 to $n$, but we are interested in evaluating the function at the left and right edges of each panel. There are $n - 1$ interior edges for the $n$ panels, and two more for the right and left boundaries of the integral. Therefore, there will be $n + 1$ edges, which can be numbered 0 through $n$.

The area of the first panel is

$$\frac{[f(0) + f(1)]\Delta x}{2}$$

or

$$\frac{[f(a) + f(1)]\Delta x}{2}$$

and the area of the last panel is

$$\frac{[f(n - 1) + f(n)]\Delta x}{2}$$

or

$$\frac{[f(n - 1) + f(b)]\Delta x}{2}$$

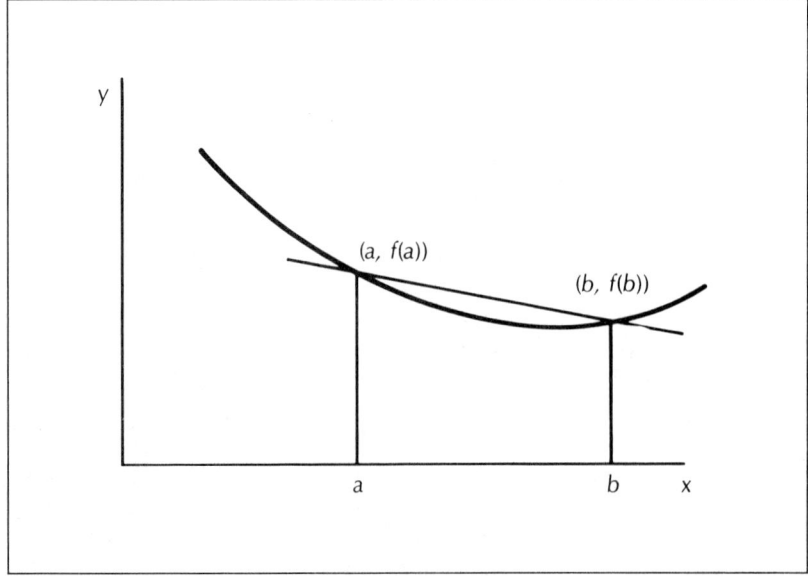

**Figure 9.2:** Calculation of the area with one panel.

The area of the *ith* panel is

$$\frac{[f(i-1) + f(i)]\Delta x}{2}$$

where $f(i-1)$ is the value of the function on the left side of the *ith* panel and $f(i)$ is the value of the function on the right side of the panel. The desired integral is the sum of the areas of all the panels. Thus the total area can be calculated by summing the areas of all the panels according to the expression

$$\{[f(a) + f(1)] + [f(1) + f(2)] + [f(2) + f(3)] + \cdots$$

$$+ [f(n-2) + f(n-1)] + [f(n-1) + f(b)]\}\frac{\Delta x}{2}$$

The right edge of the first panel is also the left edge of the second panel, and the left edge of the last panel is also the right edge of the next to last panel. The edges of all the other panels are also common to two panels. The formula for integration by the trapezoidal method can therefore be simplified as

$$[f(a) + 2f(1) + 2f(2) + \cdots +$$

$$2f(n-2) + 2f(n-1) + f(b)]\frac{\Delta x}{2}$$

Our first computer program for this method will allow us to experiment with the number of panels that we use to divide the total area.

## Programming the Trapezoidal Method

The area calculated using the trapezoidal method more closely approaches the true value as the number of panels increases. This can be demonstrated with the program given in Listing 9.1. Type the program, giving it the name TRAP1.PAS, and run it. The program asks for the number of sections into which the area is to be divided, and calculates the value based on the given number of panels.

```
PROGRAM Trap1;
{ integration by the trapezoidal method }

VAR
     Done: Boolean;
  Pieces: Integer;
  Sum, Upper, Lower: Real;

FUNCTION Fx(X: Real): Real;
{ find f(X) = 1 / X }
{ watch out for X = 0 }

BEGIN
  Fx := 1.0 / X
END;

PROCEDURE Trapez(Lower, Upper: Real;
                        Pieces: Integer;
                    VAR Sum: Real);
{ numerical integration by the trapezoidal method }
{ function is Fx, limits are Lower and Upper }
{ with number of regions equal to Pieces }
{ fixed partition is Delta_X, answer is Sum }

VAR
  I: Integer;
  X, Delta_X, Esum, Psum: Real;

BEGIN
  Delta_X := (Upper - Lower) / Pieces;
  Esum := Fx(Lower) + Fx(Upper);
  Psum := 0.0;
  FOR I := 1 TO Pieces - 1 DO
    BEGIN
      X := Lower + I*Delta_X;
      Psum := Psum + Fx(X)
    END;
  Sum := (Esum + 2.0*Psum) * Delta_X * 0.5
END; { Trapez }

BEGIN { main program }
  Done := False;
  Lower := 1.0;
  Upper := 9.0;
  WriteLn;
  REPEAT
    Write('How many sections? ');
    ReadLn(Pieces);
    IF Pieces <= 0 THEN
      Done := True
    ELSE
      BEGIN
        Trapez(Lower, Upper, Pieces, Sum);
        WriteLn(' area = ', Sum:9:5)
      END
  UNTIL Done
END.
```

**Listing 9.1:** Numerical integration by the trapezoidal method.

Procedure Trapez contains the algorithm for integration by the trapezoidal method. The function to be integrated

$$\int_1^9 \frac{dx}{x}$$

is defined in function Fx, which is a separate routine. Since this function can be directly integrated, we can compare the exact value of the integral to the value calculated by the trapezoidal method. The value of the integral is the natural logarithm of 9, or 2.197225.

We will now look at a more sophisticated version of this program.

## *Programming an Improved Trapezoidal Method*

We can improve our trapezoidal method program in two ways. First, we can change procedure Trapez so that it will automatically divide the original area into more and more pieces. Second, we can avoid much calculation at each step by using the results from the previous step.

Make a copy of the first version, using the name TRAP2.PAS. Alter the copy so that it looks like Listing 9.2.

```
PROGRAM Trap2;
{ integration by the trapezoidal method }

CONST
  Tol = 1.0E-5;

VAR
  Sum, Upper, Lower: Real;

FUNCTION Fx(X: Real): Real;

{ find f(X) = 1 / X }
{ watch out for X = 0 }

BEGIN
  Fx := 1.0 / X
END;

PROCEDURE Trapez(Lower, Upper, Tol: Real;
                       VAR Sum: Real);
{ numerical integration by the trapezoid method }
{ external function is Fx }
{ limits are Lower and Upper }
{ with number of regions equal to Pieces }
{ partition width is Delta_X, answer is Sum }
```

**Listing 9.2:** An improved trapezoidal method.

```
VAR
  Pieces, I: Integer;
  X, Delta_X, End_Sum, Mid_Sum, Sum1: Real;
BEGIN
  Pieces := 1;
  Delta_X := (Upper - Lower) / Pieces;
  End_Sum := Fx(Lower) + Fx(Upper);
  Sum := End_Sum * Delta_X / 2.0;
  WriteLn('    1', Sum:9:5);
  Mid_Sum := 0.0;
  REPEAT
    Pieces := Pieces * 2;
    Sum1 := Sum;
    Delta_X := (Upper - Lower) / Pieces;
    FOR I := 1 TO Pieces DIV 2 DO
      BEGIN
        X := Lower + Delta_X *(2.0*I - 1.0);
        Mid_Sum := Mid_Sum + Fx(X)
      END;
    Sum := (End_Sum + 2.0*Mid_Sum) * Delta_X * 0.5;
    WriteLn(Pieces:5, Sum:9:5)
  UNTIL Abs(Sum - Sum1) <= Abs(Tol * Sum)
END; { Trapez }

BEGIN { main program }
  Lower := 1.0;
  Upper := 9.0;
  WriteLn;
  Trapez(Lower, Upper, Tol, Sum);
  WriteLn;
  WriteLn(' area = ', Sum:9:5)
END.
```

**Listing 9.2:** An improved trapezoidal method (continued).

## *Running the Improved Trapezoidal Method Program*

For the first calculation, the entire area is taken as a single panel. The program then doubles the number of panels to 2 and calculates the new area. The program continues doubling the number of panels at each step. As the number of panels increases, so does the accuracy of the result (and, of course, the length of the computation time). The program displays the number of panels and the corresponding calculated areas at each step. The output looks like Figure 9.3.

The calculation process terminates when two successive values are within the desired tolerance. The tolerance is a relative global variable that is set to the value of $10^{-5}$ near the beginning of the program.

```
         1 4.44444
         2 3.02222
         4 2.46349
         8 2.27341
        16 2.21733
        32 2.20234
        64 2.19851
       128 2.19755
       256 2.19731
       512 2.19725
      1024 2.19723

    area =  2.19723
```

**Figure 9.3:** Integration by the improved trapezoidal method.

When you are satisfied that the program is working properly, you may also want to remove the WriteLn statement in procedure Trapez that displays the successive values during the iteration process.

Notice that the program does not recalculate all the panel heights at each step. Since all the panel heights from the previous step are common to the present one, it is faster to save the sum and use it for the next step. Only the heights midway between the previous points need to be computed. The program then adds these values to the previous sum of the interior values, multiplies the new sum by 2, adds the result to the values at each end of the function, and finally multiplies that result by one-half the panel width to obtain the new area.

While this improved version runs faster than the first version, it suffers from the same limitations. We are trying to fit a general curve with a set of straight lines. If the original curve is not a straight line, many steps may be needed for convergence.

Sometimes a second-order polynomial (a parabola) will be a better fit than a straight line. This is the idea of the next method that we will consider. We will study the method with three different equations: the same function that we used with the trapezoidal method program, an exponential function, and a sine function.

## Simpson's Method

*Simpson's method* of numerical integration is similar to the trapezoidal method, except that the original curve is replaced by a set of parabolas rather than a set of straight lines. Since a parabola can be defined by three points, we must divide the original area into an even number of panels. Each parabola is fitted to the tops of two adjacent panels. In general, we expect parabolas to produce a better fit than straight lines, so we should need fewer panels to obtain satisfactory convergence. The formula is

$$\left(f_0 + f_n + 4 \sum_{\substack{j=1 \\ j \text{ odd}}}^{n-1} f_j + 2 \sum_{\substack{j=2 \\ j \text{ even}}}^{n-2} f_j\right) \frac{\Delta x}{3}$$

## Programming Simpson's Method

Make a copy of the trapezoidal method program, using the name SIMP1.PAS. Change the new version to look like Listing 9.3. When you run this version the results should look like Figure 9.4.

```
PROGRAM Simp1;
{ integration by Simpson's method }

CONST
  Tol = 1.0E-5;

VAR
  Sum, Upper, Lower: Real;

FUNCTION Fx(X: Real): Real;

{ find f(X) = 1 / X }
{ watch out for X = 0 }

BEGIN
  Fx := 1.0 / X
END;   { function Fx }

PROCEDURE Simps(Lower, Upper, Tol: Real;
                       VAR Sum: Real);

{ numerical integration by Simpson's method }
{ function is Fx, limits are Lower and Upper }
{ with number of regions equal to Pieces }
{ partition is Delta_X, answer is Sum }

VAR
  I, Pieces: Integer;
  X, Delta_X, Even_Sum,
  Odd_Sum, End_Sum, Sum1: Real;
```

**Listing 9.3:** Numerical integration by Simpson's method.

```
BEGIN
  Pieces := 2;
  Delta_X := (Upper - Lower) / Pieces;
  Odd_Sum := Fx(Lower + Delta_X);
  Even_Sum := 0.0;
  End_Sum := Fx(Lower) + Fx(Upper);
  Sum := (End_Sum + 4.0*Odd_Sum) * Delta_X / 3.0;
  WriteLn(Pieces:5, Sum:9:5);
  REPEAT
    Pieces := Pieces * 2;
    Sum1 := Sum;
    Delta_X := (Upper - Lower) / Pieces;
    Even_Sum := Even_Sum + Odd_Sum;
    Odd_Sum := 0.0;
    FOR I:= 1 TO Pieces DIV 2 DO
      BEGIN
        X := Lower + Delta_X * (2.0*I - 1.0);
        Odd_Sum := Odd_Sum + Fx(X)
      END;
    Sum := (End_Sum + 4.0*Odd_Sum
         + 2.0*Even_Sum) * Delta_X / 3.0;
    WriteLn(Pieces:5, Sum:9:5)
  UNTIL Abs(Sum - Sum1) <= Abs(Tol * Sum)
END; { Simps }

BEGIN { main program }
  Lower := 1.0;
  Upper := 9.0;
  WriteLn;
  Simps(Lower, Upper, Tol, Sum);
  WriteLn;
  WriteLn(' area = ', Sum:9:5)
END.
```

**Listing 9.3:** Numerical integration by Simpson's method (continued).

```
    2 2.54815
    4 2.27725
    8 2.21005
   16 2.19864
   32 2.19734
   64 2.19723
  128 2.19723

area =  2.19723
```

**Figure 9.4:** Results of integration by Simpson's method.

Compare Figure 9.4 with Figure 9.3. At each stage, the value of the integral obtained by Simpson's method is closer to the correct answer than is the value obtained using the trapezoidal method for the same number of panels. As a result, fewer operations are needed, and thus convergence occurs more quickly.

At each step of Simpson's method, the function is evaluated only at the odd positions. The sum for the even positions is obtained from the sum of all the previous interior positions, both even and odd.

## Using an Exponential Function with the Simpson's Method Program

Let us consider a second example of integration by Simpson's method by using a function related to the normal distribution described in Chapter 2 and the error function that we will encounter in Chapter 11. Now we will integrate the curve from zero to a value of unity.

Make a copy of the previous program, using the name SIMP2.PAS. We are going to find the value of the integral

$$\int_0^1 e^{-x^2} dx$$

Therefore, change function Fx so that it looks like

```
FUNCTION Fx( X : Real) : Real;
BEGIN
   Fx := Exp( − X*X)
END;
```

Then change the integration limits for the first and second lines of the main program so that they read

```
Lower := 0.0;
Upper := 1.0;
```

Run the new version, and compare the output to Figure 9.5. Notice how rapidly the process converges for this function.

## Using a Periodic Function with the Simpson's Method Program

All the functions that we have integrated so far exhibit curvature in the same direction throughout the interval. Let us now consider the

```
                                     2    0.747180
                                     4    0.746855
                                     8    0.746826
                                    16    0.746824

                              area =    0.746824
```

**Figure 9.5:** Results of integrating an exponential function using Simpson's method.

periodic function

$$\sin^2 x$$

over the interval of 0 to 4 Π. Although the function changes back and forth from positive to negative curvature, it always has a positive value. As a consequence, the integral over any region is positive. However, this function has a zero value at both limits, at the midpoint of the interval, and at the quarter points. Therefore, the first area calculated by either the trapezoidal or Simpson's method will give a zero result. The next approximation of Simpson's method, however, will give positive results.

Make a copy of the previous program, using the name SIMP3.PAS. Alter function Fx to look like

```
BEGIN
    Fx := Sqr(Sin(X))
END;
```

Then change the limits on the first line of the main program to read

```
Lower := 0.0;
Upper := 4.0*Pi;
```

You do not have to define this symbol Pi in your program because Turbo Pascal does it for you. Run the program, and compare the output to Figure 9.6.

If we had programmed this routine to quit when the absolute value of the area was less than the value of Tol, then it would have stopped after the first iteration. The resulting calculation would have been about zero, rather than the correct value of 6.28319. Thus, we were justified in selecting a *relative* tolerance as the convergence criterion.

```
              2   0.00000
              4   0.00000
              8   8.37758
             16   6.28319
             32   6.28319

           area =    6.28319
```

**Figure 9.6:** Results of integrating a period function using Simpson's method.

In the next section, we will consider a somewhat more complicated technique for numerical integration. We will write a Turbo Pascal program, and then run the program to solve the same three functions that we examined here.

## The Romberg Method

Simpson's method, which uses a set of second-order equations, is an improvement over the trapezoidal method, in which first-order equations are used. Accordingly, we could attempt to further improve our numerical integration method by replacing the original curve with a set of cubic, or even higher order, polynomials. However, there is another approach, known as the *Romberg method*. With this technique, the area is calculated by the trapezoidal method, but the errors inherent in that method are accounted for by using interpolation.

We designate the usual sequence of the trapezoidal values by the notation $T_{11}$, $T_{21}$, $T_{31}$, etc., and assign them to the first column of the two-dimensional matrix T. We designate the first level of interpolated values as $T_{12}$, $T_{22}$, $T_{32}$, etc., and place them into column 2 of the T matrix. We can then interpolate the values in column 2 to produce a third column with values designated $T_{13}$, $T_{23}$, $T_{33}$, etc.:

$$\begin{bmatrix} T_{11} & T_{12} & T_{13} & \cdots \\ T_{21} & T_{22} & T_{23} & \cdots \\ T_{31} & T_{32} & T_{33} & \cdots \\ \cdots & \cdots & \cdots & \cdots \end{bmatrix}$$

If we continue in this way, we will find that the interpolated values rapidly converge upon the correct integral. The advantage of this method is that the function only has to be evaluated for the entries in the first column, corresponding to the regular trapezoidal rule values. Each value in the other columns is obtained from a combination of the entry directly to the left and the one just below the entry to the left. For example, to calculate

$$T_{12} = \frac{4T_{21} - T_{11}}{3}$$

$$T_{22} = \frac{4T_{31} - T_{21}}{3}$$

$$T_{13} = \frac{16T_{22} - T_{12}}{15}$$

The general algorithm is

$$T_{ij} = \frac{4^{j-1}T_{i+1,j-1} - T_{i,j-1}}{4^{j-1} - 1}$$

### *Programming the Romberg Method*

A program that does numerical integration using the Romberg method is given in Listing 9.4. Create a file named ROMB1.PAS, and type the program.

```
PROGRAM Romb1;
{ integration by the Romberg method }

CONST
  Tol = 1.0E-5;

VAR
  Done: Boolean;
  Sum, Upper, Lower: Real;

FUNCTION Fx(X: Real): Real;
{ find f(X) = 1 / X }
{ watch out for X = 0 }

BEGIN
  Fx := 1.0 / X
END;
```

**Listing 9.4:** Numerical integration by the Romberg method.

```
PROCEDURE Romb(Lower, Upper, Tol: Real;
                          VAR Ans: Real);
{ numerical integration by Romberg method }

VAR
  Nx: ARRAY[1..16] OF Integer;
  T:  ARRAY[1..136] OF Real;
  Done, Error: Boolean;
  Pieces, Nt, I, II, N, Nn,
  L, Ntra, K, M, J: Integer;
  Delta_X, C, Sum, Fotom, X: Real;

BEGIN
  Done := False;
  Error := False;
  Pieces := 1;
  Nx[1] := 1;
  Delta_X := (Upper-Lower)/Pieces;
  C := (Fx(Lower)+Fx(Upper))* 0.5;
  T[1] := Delta_X*C;
  N := 1;
  Nn := 2;
  Sum := C;

  REPEAT
    N := N+1;
    Fotom := 4.0;
    Nx[N] := Nn;
    Pieces := Pieces * 2;
    L := Pieces - 1;
    Delta_X := (Upper-Lower)/Pieces;
    { compute trapezoidal Sum for 2^(N-1)+1 points }
    FOR II := 1 TO (L+1) DIV 2 DO
      BEGIN
        I := II * 2 - 1;
        X := Lower + I * Delta_X;
        Sum := Sum + Fx(X)
      END;
    T[Nn] := Delta_X*Sum;
    Write(Pieces:5, T[Nn]:9:5);
    Ntra := Nx[N-1];
    K := N-1;
    { compute N-th row of T array }
    FOR M := 1 TO K DO
      BEGIN
        J := Nn+M;
        Nt := Nx[N-1]+M-1;
        T[J] := (Fotom*T[J-1] - T[Nt])/(Fotom-1.0);
        Fotom := Fotom*4.0
      END;
    WriteLn(J:4, T[J]:9:5);
    IF N > 4 THEN
```

**Listing 9.4:** Numerical integration by the Romberg method (continued).

```
        BEGIN
          IF T[Nn+1] <> 0.0 THEN
              IF (Abs(T[Ntra+1] - T[Nn+1]) <= Abs(T[Nn+1]*Tol))
                OR (Abs(T[Nn - 1] - T[J]) <= Abs(T[J]*Tol)) THEN
                    Done := True
              ELSE IF N > 15 THEN
                BEGIN
                    Done := True;
                    Error := True
                END
        END; { IF N > 4 }
      Nn := J+1
    UNTIL Done;
    Ans := T[J]
END; { Romberg }

BEGIN { main program }
  Lower := 1.0;
  Upper := 9.0;
  WriteLn;
  Romb(Lower, Upper, Tol, Sum);
  WriteLn;
  WriteLn(' area = ', Sum:9:5)
END.
```

**Listing 9.4:** Numerical integration by the Romberg method (continued).

### *First Run of the Romberg Method Program*

When you run the Romberg method program, it displays the regular trapezoidal values and the interpolated values at each step. The results look like Figure 9.7.

By comparing the values shown in Figure 9.7 with those shown in the previous figures, you can see that in this case the Romberg method converges even more rapidly than Simpson's method.

### *Using an Exponential Function with the Romberg Method Program*

Copy the Romberg method program, using the name ROMB2.PAS. Alter the copy so that it calculates the integral

$$\int_0^1 e^{-x^2} dx$$

Change function Fx so that it reads

**FUNCTION** Fx(X: Real): Real;

```
BEGIN
   Fx := Exp(−X*X)
END;
```

and change the limits in the first line in the main program to read:

```
Lower := 0.0;
Upper := 1.0;
```

Run the Romberg method program with the new formula. The results should look like Figure 9.8. In this example, convergence is not as rapid as it was with Simpson's method.

```
             2 3.02222   3 2.54815
             4 2.46349   6 2.25919
             8 2.27341  10 2.20472
            16 2.21733  15 2.19773
            32 2.20234  21 2.19724
            64 2.19851  28 2.19723

          area =  2.19723
```

**Figure 9.7:** Results of integration by the Romberg method.

```
             2  0.731370   3  0.747180
             4  0.742984   6  0.746834
             8  0.745866  10  0.746824
            16  0.746585  15  0.746824

          area =  0.746824
```

**Figure 9.8:** Results of integrating an exponential function using the Romberg method.

## Using a Periodic Function with the Romberg Method Program

We now return to the integral

$$\int_0^{4\pi} \sin^2 x \, dx$$

that we considered earlier. If we attempt to solve this equation using the Romberg method, we will obtain a value of zero for the first two approximations. However, if we are careful to use a relative tolerance rather than an absolute one to determine convergence, we will find the correct solution after several more iterations.

Make a copy of the previous program, using the name ROMB3.PAS. Change function Fx so that it reads

```
BEGIN
    Fx := Sqr(Sin(X))
END;
```

and the first and second lines of the main program become

```
Lower := 0.0;
Upper := 4.0*pi;
```

Run the program and compare the results to Figure 9.9.

```
          2 0.00000      3 0.00000
          4 0.00000      6 0.00000
          8 6.28319     10 9.07793
         16 6.28319     15 6.08755
         32 6.28319     21 6.28633

        area =  6.28633
```

**Figure 9.9:** Results of integrating a periodic function using the Romberg method.

In the next section, we will expand our Romberg method program to deal with a special case—a function that approaches infinity at one of the limits to the area beneath the curve. To solve this problem, we will develop a technique of adjusting the width of the panels as they approach the infinite limit until we arrive at a sufficiently accurate value for the total area.

# Solving Functions
# that Become Infinite at One Limit

For each of the above methods, we have used a uniform panel width throughout the desired interval. However, more panels are required when the function has a greater slope. Conversely, fewer panels can be used when the slope is smaller. Thus, we can reduce the number of calculations by adjusting the panel width according to the slope of the function. As an extension of this idea, consider a function that approaches infinity at one limit even though the area for the interval is finite. An example is the integral of the reciprocal of the square root of x over the range from zero to unity:

$$\int_0^1 \frac{dx}{\sqrt{x}}$$

Direct integration gives an exact value of 2.0. However, if we attempt to solve this function using one of the previous integration programs, we immediately run into a problem. The value of f(a) at the left edge is infinite, so we must choose some other lower limit. If we choose a small lower limit such as $10^{-11}$, convergence will take many iterations. If we choose a larger value, such as 0.01, for the left limit, the result will be inaccurate.

## Programming Adjustable
## Panels for an Infinite Function

One method of integrating an infinite function is to start the integration at a value somewhat larger than zero. We will initially choose the left boundary to be 0.1 and evaluate the integral over the range 0.1 to 1.0. To test the reasonableness of this, we should then evaluate the next region to the left, from 0.01 to 0.1. If this area is also significant, then we must add its area to the area obtained for the first regions. We then take the next region, from 0.001 to 0.01, and see how large it is. In this way, we can take regions closer and closer to zero, observing the additional area as we progress. The program shown in Listing 9.5 uses this approach with the Romberg method.

```
PROGRAM Romb3;
{ integration by the Romberg method }

CONST
  Tol = 1.0E-5;

VAR
  Done: Boolean;
  SumT, Sum, Upper, Lower: Real;

FUNCTION Fx(X: Real): Real;
{ find f(X) = 1 / Sqrt(X) }
{ watch out for X = 0 }

BEGIN
  Fx := 1.0 / Sqrt(X)
END;

PROCEDURE Romb(Lower, Upper, Tol: Real;
                      VAR Ans: Real);
{ numerical integration by Romberg method }

VAR
  Nx: ARRAY[1..16] OF Integer;
  T:  ARRAY[1..136] OF Real;
  Done, Error: Boolean;
  Pieces, Nt,I, II, N, Nn,
  L, Ntra, K, M, J: Integer;
  Delta_X, C, Sum, Fotom, X: Real;

BEGIN
  Done := False;
  Error := False;
  Pieces := 1;
  Nx[1] := 1;
  Delta_X := (Upper-Lower)/Pieces;
  C := (Fx(Lower)+Fx(Upper))* 0.5;
  T[1] := Delta_X*C;
  N := 1;
  Nn := 2;
  Sum := C;
  REPEAT
    N := N+1;
    Fotom := 4.0;
    Nx[N] := Nn;
    Pieces := Pieces * 2;
    L := Pieces - 1;
    Delta_X := (Upper-Lower)/Pieces;
```

**Listing 9.5:** Programming the Romberg method with adjustable panels.

```
{ compute trapezoidal sum for 2^(N-1)+1 points }
FOR II := 1 TO (L+1) DIV 2 DO
   BEGIN
      I := II * 2 - 1;
      X := Lower + I*Delta_X;
      Sum := Sum + Fx(X)
   END;
T[Nn] := Delta_X*Sum;
Ntra := Nx[N-1];
K := N-1;
{ compute N-th row of T array }
FOR M := 1 TO K DO
   BEGIN
      J := Nn+M;
      Nt := Nx[N-1]+M-1;
      T[J] := (Fotom*T[J-1] - T[Nt])/(Fotom-1.0);
      Fotom := Fotom*4.0
   END;
IF N > 4 THEN
   BEGIN
      IF T[Nn+1] <> 0.0 THEN
         IF (Abs(T[Ntra+1] - T[Nn+1]) <= Abs(T[Nn+1]*Tol))
         OR (Abs(T[Nn - 1] - T[J]) <= Abs(T[J]*Tol)) THEN
            Done := True
         ELSE IF N > 15 THEN
            BEGIN
               Done := True;
               Error := True
            END
   END; { IF N > 4 }
Nn := J+1
UNTIL Done;
Ans := T[J]
END; { Romberg }

BEGIN { main program }
   Lower := 0.1;
   Upper := 1.0;
   WriteLn;
   SumT := 0.0;
   WriteLn(' new area    total area   lower',
      '    upper  limits');
   REPEAT
      Romb(Lower, Upper, Tol, Sum);
      Upper := Lower;
      Lower := 0.1 * Upper;
      SumT := SumT + Sum;
      WriteLn(Sum:9:6,'    ', SumT:9:5,
         '    ',Lower:12, '    ', Upper:12)
   UNTIL Abs(Sum) < Tol
END.
```

**Listing 9.5:** Programming the Romberg method with adjustable panels (continued).

### *Running the Adjustable Panel Program*

Make a copy of the previous program, giving it the name ROMB4.PAS. Change this version to look like Listing 9.5.

We have removed the WriteLn statement in procedure Romb, so that the program displays only the final value of each major area. The main program repeatedly calls procedure Romb with limits that are closer and closer to zero. When the new area is less than the tolerance, the procedure is terminated. The results should look like Figure 9.10.

If we did not remove the intermediate WriteLn statements in Romb, we would see that each of the regions was divided into 64 panels. The width of each region is one-tenth that of the region immediately to the right, and the final left limit is $10^{-13}$. Thus, if the whole region from $10^{-13}$ to 1.0 were integrated as one unit, it would require over a million million uniformly spaced panels. Such a method would take about one thousand million times longer to produce the correct answer.

```
new area    total area   lower    upper   limits
1.367545    1.36754   1.00000E-002   1.00000E-001
0.432456    1.80000   1.00000E-003   1.00000E-002
0.136754    1.93675   1.00000E-004   1.00000E-003
0.043246    1.98000   1.00000E-005   1.00000E-004
0.013675    1.99368   1.00000E-006   1.00000E-005
0.004325    1.99800   1.00000E-007   1.00000E-006
0.001368    1.99937   1.00000E-008   1.00000E-007
0.000432    1.99980   1.00000E-009   1.00000E-008
0.000137    1.99994   1.00000E-010   1.00000E-009
0.000043    1.99998   1.00000E-011   1.00000E-010
0.000014    1.99999   1.00000E-012   1.00000E-011
0.000004    2.00000   1.00000E-013   1.00000E-012
```

**Figure 9.10:** Integration of a function that becomes infinite at one limit.

## *Summary*

In this chapter, we developed and compared three different numerical integration programs. We started with the trapezoidal and Simpson's methods, and progressed to the more sophisticated Romberg method, which uses a matrix of interpolations. We tried these methods using several different functions, and we used a refined Romberg method program to compute the area under the curve of an infinite function.

All the above integrations, except the last, were performed on analytic functions with uniformly spaced panels. Other techniques, which are not discussed here, are required with discrete data. One approach is to fit the data to a polynomial function, using one of the techniques discussed in earlier chapters, then integrate the resulting polynomial.

# *Exercises*

**9-1.** Find the value of the integral

$$\int_0^{10} x^3 e^{-x}\, dx$$

using the trapezoidal, Simpson's, and Romberg methods. Which method seems best?

**9-2.** Find the value of the integral

$$\int_0^{10} \frac{x\, dx}{1 + e^x}$$

using the trapezoidal, Simpson's, and Romberg methods. Which method seems best?

**9-3.** Find the value of the integral

$$\int_0^{10} \frac{dx}{e^x + e^{-x}}$$

using the trapezoidal, Simpson's, and Romberg methods. Which method seems best?

**9-4.** Find the value of the integral

$$\int_0^{10} \frac{dx}{(1 + x^2)^2}$$

using the trapezoidal, Simpson's, and Romberg methods. Which method seems best?

# Nonlinear Curve-Fitting Equations

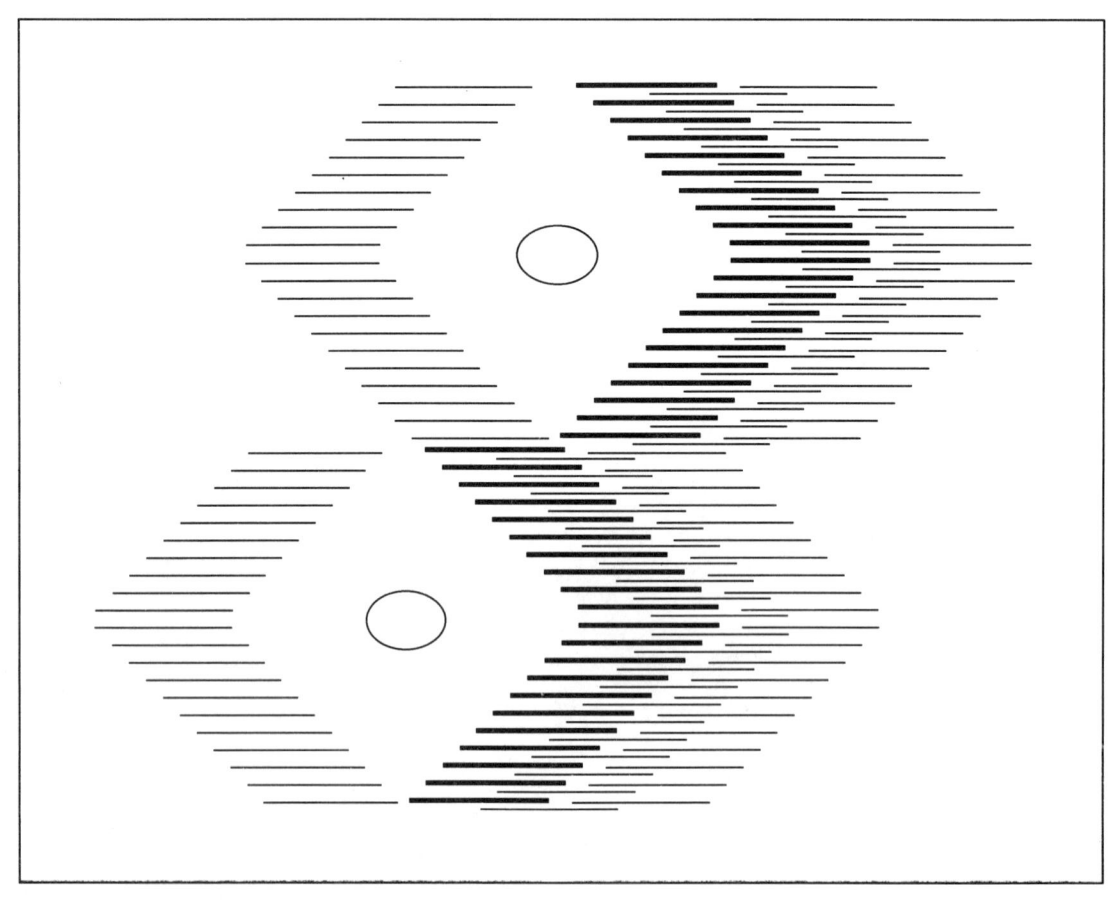

*Chapter* **10**

In Chapters 5 and 6, we developed computer programs for finding the coefficients to various curve-fitting equations. Since we chose approximating functions with linear coefficients, the resulting equations were linear, and therefore easily solved. Sometimes, however, it is necessary to choose approximating functions with nonlinear coefficients. Then, the calculation of the coefficients can be more difficult.

In this chapter, we consider two different techniques for nonlinear curve fitting. In one method, we linearize the approximating function, and then find the solution to the linear form. Using this first method, we will fit curves for two different sets of data. First, we fit a linearized form of the so-called rational function to data representing the Clausing factor. Then, we will fit a linearized exponential function to data describing the diffusion of zinc in copper over a given temperature range.

Our second approach to nonlinear curve fitting is more direct. With this method, we eliminate one coefficient and then solve the resulting equations by using Newton's method, which we studied in Chapter 8.

The approaches discussed in this chapter are not *general* methods for nonlinear curve fitting. There is no universal approach to such solutions. We will explore techniques that can be applied to specific equations.

Let us begin with our first example of the linearization method.

# Linearizing the Rational Function

A commonly used, nonlinear approximating function is known as the *rational function*. This expression is formed from the ratio of two polynomials

$$y = \frac{A_1 + A_3x + A_5x^2 \ldots}{1 + A_2x + A_4x^2 + A_6x^3 \ldots}$$

In this expression, x is the independent variable, y is the dependent variable, and $A_1$, $A_2$, etc., are the coefficients, as usual.

We can make the nonlinear rational function linear by the following operations. Multiply both sides of the equation by the denominator polynomial:

$$y(1 + A_2x + A_4x^2 + A_6x^3 \ldots) = A_1 + A_3x + A_5x^2 \ldots$$

Rearrange the terms of the new equation:

$$y = A_1 - A_2xy + A_3x - A_4x^2y + A_5x^2 - A_6x^3y \ldots$$

Some of the terms on the right contain the dependent variable, y, as well as the independent variable, x. But remember, both x and y are arrays of known values. It is the unknown coefficients $A_1$, $A_2$, .. $A_n$ that we want to obtain. All these coefficients are now linear, so we can determine them by using one of the methods developed in Chapters 5 and 7.

# Fitting the Clausing Factor to the Rational Function

A program for obtaining a least-squares fit to the linearized form of the rational function is given in Listing 10.1. Create a file named FITPOL.PAS, and copy the program given in Listing 7.2. Change the program to look like Listing 10.1.

The data in this program represent the *Clausing factor* as a function of length to radius (L/r) for cylindrical orifices. When molecules with a long mean free path effuse through a cylindrical orifice, some of them strike the orifice walls and are returned in the opposite direction. The remaining molecules continue through to the other side of the orifice. The Clausing factor gives the fraction of those molecules that emerge from the other end. The Clausing factor ranges from zero to unity and becomes smaller as the cylinder increases in length or decreases in radius. Thus, the ratio L/r appears in a formula for the direct calculation of the Clausing factor. We have simplified the equation by substituting x for L/r.

```
PROGRAM Fitpol;
{ Turbo Pascal program to perform a
  linear least-squares fit
  to the ratio of two polynomials.
  Separate procedure Square (Listing 3.2) and
  Gaussj (Listing 4.4) needed }

CONST
  Maxr = 20; { data points }
  Maxc = 4; { polynomial terms }

TYPE
  Ary   = ARRAY[1..Maxr] OF Real;
  Arys  = ARRAY[1..Maxc] OF Real;
  Ary2  = ARRAY[1..Maxr, 1..Maxc] OF Real;
  Ary2s = ARRAY[1..Maxc, 1..Maxc] OF Real;

VAR
  X, Y, Y_Calc, Resid: Ary;
    Coef, Sig: Arys;
   Nrow, Ncol: Integer;
  Correl_Coef: Real;

PROCEDURE Get_Data
           (VAR X    : Ary; { independent variable }
            VAR Y    : Ary; { dependent variable }
            VAR Nrow: Integer); { length of vectors }

VAR
  I : Integer;

BEGIN
  { Clausing Factors }
  Nrow := 10;
  X[1]:= 0.1;  Y[1]:=  0.9524;
  X[2]:= 0.2;  Y[2]:=  0.9092;
  X[3]:= 0.5;  Y[3]:=  0.8013;
  X[4]:= 1.0;  Y[4]:=  0.6720;
  X[5]:= 1.2;  Y[5]:=  0.6322;
  X[6]:= 1.5;  Y[6]:=  0.5815;
  X[7]:= 2.0;  Y[7]:=  0.5142;
  X[8]:= 3.0;  Y[8]:=  0.4201;
  X[9]:= 4.0;  Y[9]:=  0.3566;
  X[10]:= 6.0; Y[10]:= 0.2755;
END; { procedure Get_Data }

PROCEDURE Write_Data;
{ display the answers }

VAR
  I: Integer;
```

**Listing 10.1:** Fitting the Clausing factor to the ratio of two polynomials.

```
BEGIN
  WriteLn;
  WriteLn;
  WriteLn('  I      X        Y       Y CALC      RESID');
  FOR I := 1 TO Nrow DO
    WriteLn(I:3, X[I]:8:1, Y[I]:9:4,
      Y_Calc[I]:9:4, Resid[I]:9:4);
  WriteLn;
  WriteLn('coefficients     errors');
  WriteLn(Coef[1]:8:5, '      ', Sig[1]:13, ' Constant term');
  FOR I := 2 TO Ncol DO
    WriteLn
      (Coef[I]:8:5, '      ', Sig[I]:13);      { other terms }
  WriteLn;
  WriteLn(' Correlation coefficient is ', Correl_Coef:8:5)
END; { Write_Data }

{$I SQUARE.PAS }
{$I GAUSSJ.PAS }

PROCEDURE Linfit(X,        { independent variable }
             Y: Ary;  { dependent variable }
         VAR Y_Calc: Ary; { calculated dependent variable }
         VAR Resid : Ary; { array of residuals }
         VAR Coef : Arys; { coefficients }
         VAR Sig  : Arys; { errors on coefficients }
             Nrow : Integer; { length of Ary }
         VAR Ncol : Integer); { number of terms }

{ least-squares fit to }
{ Nrow sets of X and Y pairs of points }
{ Separate procedures needed:
    Square - form square coefficient matrix
    Gaussj - Gauss-Jordan elimination      }

VAR
  Xmatr    : Ary2;  { Data matrix }
  A        : Ary2s; { coefficient matrix }
  G        : Arys;  { constant vector }
  Error    : Boolean;
  I, J, Nm: Integer;
  Xi, Yi, Yc, SRS, SEE,
  Sum_Y, Sum_Y2: Real;

BEGIN                { procedure Linfit }
  Ncol := 4; { number of terms }
  FOR I := 1 TO Nrow DO
    BEGIN    { setup X matrix }
      Xi := X[I];
      Yi := Y[I];
      Xmatr[I, 1] := 1.0; { first column }
      Xmatr[I, 2] := -Xi*Yi;
      Xmatr[I, 3] := Xi;
      Xmatr[I, 4] := -Sqr(Xi)*Yi
    END;
```

**Listing 10.1:** Fitting the Clausing factor to the ratio of two polynomials (continued).

```
      Square(Xmatr, Y, A, G, Nrow, Ncol);
      Gaussj(A, G, Coef, Ncol, Error);
      Sum_Y := 0.0;
      Sum_Y2 := 0.0;
      SRS := 0.0;
      FOR I := 1 TO Nrow DO
        BEGIN
          Xi := X[I];
          Yi := Y[I];
          Yc := Coef[1]
            +(-Coef[2]*Yi+Coef[3]-Coef[4]*Xi*Yi)*Xi;
          Y_Calc[I] := Yc;
          Resid[I] := Yc - Yi;
          SRS := SRS + Sqr(Resid[I]);
          Sum_Y := Sum_Y + Yi;
          Sum_Y2 := Sum_Y2 + Yi * Yi
        END;
      Correl_Coef :=  Sqrt(1.0 - SRS /
            (Sum_Y2 - Sqr(Sum_Y) / Nrow));
      IF Nrow = Ncol THEN
        Nm := 1
      ELSE
        Nm := Nrow - Ncol;
      SEE := Sqrt(SRS / Nm);
      FOR I := 1 TO Ncol DO    { errors in solution }
        Sig[I] := SEE * Sqrt(A[I, I])
    END; { Linfit }

    BEGIN  { main program }
      Get_Data(X, Y, Nrow);
      Linfit(X, Y, Y_Calc, Resid, Coef, Sig, Nrow, Ncol);
      Write_Data
    END.
```

**Listing 10.1:** Fitting the Clausing factor to the ratio of two polynomials (continued).

## *Running the Clausing Factor Fitting Program*

The program calculates the data matrix for four terms, corresponding to a first-order numerator and a second-order denominator. You can add terms to the approximating function, but you must change the symbols Maxc and Ncol to reflect the desired number of terms. In addition, you must place expressions such as

```
Matr[I,5] := Matr[I,3]*XI;
Matr[I,6] := Matr[I,4]*XI;
```

in procedure Linfit. The results shown in Figure 10.1 correspond to the equation

$$Y = \frac{1.0017 + 0.237x}{1 + 0.7522x + 0.0912x^2}$$

If the data are fitted with a regular polynomial function, rather than with the rational function, the result will not be as good (for the same number of coefficients in the approximating function).

In our next example of the linearized approach, we examine an exponential equation. Later in this chapter, we will fit the same equation using another, more direct approach.

```
    I      X      Y     Y CALC    RESID
    1     0.1   0.9524   0.9529   0.0005
    2     0.2   0.9092   0.9090  -0.0002
    3     0.5   0.8013   0.8006  -0.0007
    4     1.0   0.6720   0.6719  -0.0001
    5     1.2   0.6322   0.6324   0.0002
    6     1.5   0.5815   0.5818   0.0003
    7     2.0   0.5142   0.5146   0.0004
    8     3.0   0.4201   0.4199  -0.0002
    9     4.0   0.3566   0.3563  -0.0003
   10     6.0   0.2755   0.2756   0.0001

  coefficients      errors
    1.00169      4.602150E-004  Constant term
    0.75215      1.370260E-002
    0.23700      1.219420E-002
    0.09122      5.143180E-003

  Correlation coefficient is  1.00000
```

**Figure 10.1:** The Clausing factor versus L/r fitted to a rational function.

# Linearizing the Exponential Equation

A common nonlinear equation has the form

$$y = Ae^{Bx}$$

where x is the independent variable, y is the dependent variable, and A and B are the desired coefficients. This equation is widely used in scientific and engineering calculations because it is the solution to a first-order differential equation.

The equation can be linearized by taking the logarithm. The result is

$$\ln y = \ln A + Bx$$

In this form, the dependent variable is ln y and the unknown coefficients are ln A and B. Since the new coefficients are linear, we can obtain a least-squares fit by using the procedure Linfit, which we developed in Chapter 5.

## Programming an Exponential Curve Fit

Listing 10.2 gives a program for finding a least-squares fit to the linearized exponential equation. Copy Listing 7.2, using the name DIFFUS.PAS. As shown in Listing 10.2, change the value of Ncol to 2, and remove the calculation of the standard error.

The data given in procedure Get_Data represent the diffusion of zinc in copper over the temperature range 600°C to 900°C. The diffusion equation is

$$D = D_0 e^{-Q/RT}$$

where D is the diffusion coefficient in $cm^2/sec$, $D_0$ is the diffusion constant in the same units, Q is the activation energy in cal/mole, R is the gas constant, and t is the temperature in kelvins. The independent variable, x, is the reciprocal of the temperature in kelvins. The dependent variable y, is the logarithm of the diffusion coefficient.

```
PROGRAM Diffus;
{ Turbo Pascal program to perform a
  linear least-squares fit
  for the diffusion of Zn in Cu.
  Separate procedure Square (Listing 3.2) and
  Gaussj (Listing 4.4) needed }

CONST
  Maxr = 20;    { data points }
  Maxc = 4;     { polynomial terms }
  R    = 1.987; { gas constant }

TYPE
  Ary   = ARRAY[1..Maxr] OF Real;
  Arys  = ARRAY[1..Maxc] OF Real;
  Ary2  = ARRAY[1..Maxr, 1..Maxc] OF Real;
  Ary2s = ARRAY[1..Maxc, 1..Maxc] OF Real;
```

**Listing 10.2:** A least-squares fit to the linearized exponential equation.

```
VAR
   X, Y, Y_Calc, T, D, Resid: Ary;
         Coef, Sig: Arys;
         Nrow, Ncol: Integer;
   Correl_Coef, SRS: Real;

PROCEDURE Get_Data(VAR X, Y, T, D: Ary;
                        VAR Nrow: Integer);
{ Get values for Nrow and arrays T, D }

VAR
   I: Integer;

BEGIN
   Nrow := 7;
   T[1]:=   600.0; D[1]:= 1.4E-12;
   T[2]:=   650.0; D[2]:= 5.5E-12;
   T[3]:=   700.0; D[3]:= 1.8E-11;
   T[4]:=   750.0; D[4]:= 6.1E-11;
   T[5]:=   800.0; D[5]:= 1.6E-10;
   T[6]:=   850.0; D[6]:= 4.4E-10;
   T[7]:=   900.0; D[7]:= 1.2E-9;
   FOR I := 1 TO Nrow DO
     BEGIN
        X[I] := 1.0/(T[I] + 273.0);
        Y[I] := Ln(D[I])
     END
END; { procedure Get_Data }


PROCEDURE Write_Data;
{ display the answers }

VAR
   I: Integer;

BEGIN
   WriteLn;
   WriteLn;
   WriteLn('  I   T C            D           ',
      '   D Calc');
   FOR I := 1 TO Nrow DO
     WriteLn(I:3, T[I]:6:0,'  ',D[I]:12, '   ',
        Y_Calc[I]:12);
   WriteLn;
   WriteLn('coefficients');
   WriteLn(Coef[1]:12, '       Constant term');
   FOR I := 2 TO Ncol DO
     WriteLn
        (Coef[I]:12);      { other terms }
   WriteLn;
   WriteLn(' D0 = ', (Exp(Coef[1])):7:2,' cm sq/sec.');
   WriteLn(' Q = ', (-R * Coef[2] / 1000.0):8:2,
      ' kcal/mole');
   WriteLn; WriteLn(' SRS = ', SRS:7:3)
END; { Write_Data }
```

**Listing 10.2:** A least-squares fit to the linearized exponential equation (continued).

```
{$I Square.PAS }
{$I Gaussj.PAS }

PROCEDURE Linfit(X,         { independent variable }
                 Y: Ary; { dependent variable }
         VAR Y_Calc: Ary; { calculated dependent variable }
         VAR Resid : Ary; { array of residuals }
         VAR Coef  : Arys; { coefficients }
         VAR Sig   : Arys; { errors on coefficients }
             Nrow  : Integer; { length of Ary }
         VAR Ncol  : Integer); { number of terms }

{ least-squares fit to }
{ Nrow sets of X and Y pairs of points }
{ Separate procedures needed:
   Square - form Square coefficient matrix
   Gaussj - Gauss-Jordan elimination       }

VAR
  Xmatr  : Ary2; { Data matrix }
  A      : Ary2s; { coefficient matrix }
  G      : Arys; { constant vector }
  Error  : Boolean;
  I, J, Nm: Integer;
  SEE, A1 : Real;

BEGIN                { procedure Linfit }
  Ncol := 2; { number of terms }
  FOR I := 1 TO Nrow DO
    BEGIN    { set up X matrix }
      Xmatr[I, 1] := 1.0; { first column }
      Xmatr[I, 2] := X[I] { second column }
    END;
  Square(Xmatr, Y, A, G, Nrow, Ncol);
  Gaussj(A, G, Coef, Ncol, Error);
  SRS := 0.0;
  A1 := Exp(Coef[1]);
  FOR I := 1 TO Nrow DO
    BEGIN
      Y_Calc[I] := A1 * Exp(Coef[2] * X[I]);
      IF Y[I] <> 0.0 THEN
        Resid[I] := Y_Calc[I] / Y[I] - 1.0
      ELSE
        Resid[I] := Y[I] / Y_Calc[I] - 1.0;
      SRS := SRS + Sqr(Resid[I])
    END
END; { Linfit }

BEGIN           { main program }
  Get_Data(X, Y, T, D, Nrow);
  Linfit(X, Y, Y_Calc, Resid, Coef, Sig, Nrow, Ncol);
  Write_Data
END.
```

**Listing 10.2:** A least-squares fit to the linearized exponential equation (continued).

### Running the Exponential Curve-Fitting Program

Run the program, and compare the results with Figure 10.2. The diffusion constant ($D_0$) can be calculated from the antilogarithm (the exponent) of the first coefficient:

$D_0 := Exp(Coef[1])$

Multiplying the second coefficient ($Coef[2]$) by the gas constant and changing the sign gives the activation energy Q:

$Q := -R * Coef[2]$

In this example, the sum of residuals squared (SRS) is given instead of the usual standard error of the coefficients. The nonlinear transform of the approximating function makes these sigmas meaningless. The calculation of this form of SRS is discussed more fully in the next section, where we will develop our second approach to nonlinear curve fitting.

```
        I    T C        D           D Calc
        1    600    1.40000E-012   1.31283E-012
        2    650    5.50000E-012   5.43313E-012
        3    700    1.80000E-011   1.94311E-011
        4    750    6.10000E-011   6.13539E-011
        5    800    1.60000E-010   1.74041E-010
        6    850    4.40000E-010   4.49923E-010
        7    900    1.20000E-009   1.07266E-009

        coefficients
        -1.1395E+000         Constant term
        -2.2889E+004

        D0 =      0.32 cm sq/sec.
        Q =      45.48 kcal/mole

        SRS =    7.000
```

**Figure 10.2:** The diffusion of zinc in copper (linearized fit).

## Direct Solution of the Exponential Equation

In the previous section, we linearized the exponential equation:

$$y = Ae^{Bx}$$

by taking the logarithm

$$\ln y = \ln A + Bx$$

We then made a least-squares fit to this linearized form of the equation. However the coefficients that produce the minimum SRS to the linearized form will not, in general, produce the minimum SRS for the original, nonlinearized equation. Thus, we should consider a direct, least-squares solution to the nonlinear equation.

For our direct approach to solving the exponential function, we will follow the usual curve-fitting algorithm up to the point of taking the derivatives of the SRS. Then we will see a way to eliminate one coefficient from the equation, and we will solve the resulting expression using Newton's method.

## *Calculating the SRS*

The residuals for the equation

$$y = Ae^{Bx}$$

can be defined as

$$r = Ae^{Bx} - y$$

However, if the data are all measured to about the same relative degree of precision, independent of the magnitude, it is more meaningful to use a relative residual:

$$r = (Ae^{Bx} - y)/y$$

or

$$r = (Ae^{Bx}/y) - 1$$

The residuals in the new form are squared, and then summed to form the SRS. The derivative of the resulting SRS is taken with respect to both A and B, and the resulting equations are set to zero. This is the same approach that we used to solve the linear equation. There are two equations and two unknowns:

$$A\sum(e^{2Bx}/y^2) - \sum(e^{Bx}/y) = 0$$

and

$$A\sum(xe^{2Bx}/y^2) - \sum(xe^{Bx}/y) = 0$$

Now, however, the resulting equations are nonlinear, and so we cannot solve them with procedure Linfit.

## Eliminating Coefficient A

Fortunately, coefficient A is linear in this example, and we can separate it. We can rearrange the first of the two above equations to give

$$A = \sum (e^{Bx}/y) / \sum (e^{2Bx}/y^2)$$

Then we substitute the following expression for A into the second equation to give

$$\sum (e^{Bx}/y) \sum (xe^{2Bx}/y^2)$$

$$- \sum (e^{2Bx}/y^2) \sum (xe^{Bx}/y) = 0 = f(B)$$

Since we have eliminated A, this equation is only a function of B. In Chapter 8, we developed a program to solve nonlinear equations by Newton's method. We can use this method here.

## Applying Newton's Method

We next take the derivative of the above equation with respect to B. The result is

$$f'(B) = 2 \sum (e^{Bx}/y) \sum (x^2 e^{2Bx}/y^2)$$

$$- \sum (xe^{2Bx}/y^2) \sum (xe^{Bx}/y)$$

$$- \sum (e^{2Bx}/y^2) \sum (x^2 e^{Bx}/y)$$

Before going any further, check to see that the equations for f(B) and f'(B) are homogeneous. The units of each term of f(B) must correspond to

$$\frac{x}{y^3}$$

and the units of $f'(B)$ must correspond to

$$\frac{x^2}{y^3}$$

## *Programming a Nonlinearized Exponential Curve Fit*

The program shown in Listing 10.3 can be used to find a nonlinear least-squares curve fit to the diffusion equation that we considered in the previous section. Since we are using Newton's method, we need a first approximation for the value of B. We can obtain a good first value from the linear equation for B, but instead of performing the complete linearized fit, we can simply calculate the value of B from Equation 20 of Chapter 5. Here, the value of y is replaced by Ln y.

$$B = \frac{\Sigma[x\ln(y)] - \Sigma(x)\,\Sigma[\ln(y)/n]}{\Sigma x^2 - (\Sigma x)^2/n}$$

The instructions for this first approximation are included in procedure Nlin. A more complicated approach is to call procedure Gauss (Listing 4.4) for the first approximation.

```
PROGRAM Nlin3;
{ Turbo Pascal program to perform a
  nonlinear least-squares fit
  for the diffusion of Zn in Cu }

CONST
  Maxr = 20;     { data points }
  Maxc = 4;      { polynomial terms }
  R    = 1.987;  { gas constant }

TYPE
  Index = 1..Maxr;
  Ary   = ARRAY[Index] OF Real;
  Arys  = ARRAY[1..Maxc]  OF Real;
  Ary2  = ARRAY[1..Maxr, 1..Maxc] OF Real;

VAR
  X, Y, Y_Calc, T, D, Ex: Ary;
  Coef: Arys; { solution vector }
  I, N, Nrow, Ncol: Integer;
  Done, Error: Boolean;
  Correl_Coef, A, B, X2, SRS: Real;
```

**Listing 10.3:** A least-squares fit to the nonlinearized exponential function.

```
PROCEDURE Get_Data(VAR X, Y: Ary;
                   VAR N: Integer);
{ Get values for N and arrays T, D }

VAR
  I: Integer;

BEGIN
  N := 7;
  T[1]:=  600.0;  D[1]:= 1.4E-12;
  T[2]:=  650.0;  D[2]:= 5.5E-12;
  T[3]:=  700.0;  D[3]:= 1.8E-11;
  T[4]:=  750.0;  D[4]:= 6.1E-11;
  T[5]:=  800.0;  D[5]:= 1.6E-10;
  T[6]:=  850.0;  D[6]:= 4.4E-10;
  T[7]:=  900.0;  D[7]:= 1.2E-9;
  FOR I := 1 TO N DO
    BEGIN
      X[I] := 1.0/(T[I] + 273);
      Y[I] := D[I]
    END
END; { procedure Get_Data }

PROCEDURE Write_Data;
{ display the answers }

VAR
  I: Integer;

BEGIN
  WriteLn;
  WriteLn(' I    T C         D ',
         '                D Calc');
  FOR I := 1 TO N DO
    WriteLn(I:3, T[I]:6:0,'  ',D[I]:12,
           '  ', Y_Calc[I]:12);
  WriteLn;
  WriteLn(' Coefficients');
  WriteLn(Coef[1]:12, '     Constant term');
  FOR I := 2 TO Ncol DO
    WriteLn(Coef[I]:12); { other terms }
  WriteLn;
  WriteLn(' DO = ', A:7:2,' cm sq/sec.');
  WriteLn(' Q = ',(-R*B/1000):8:2, ' kcal/mole');
  WriteLn; WriteLn(' SRS = ', SRS:8:4)
END; { Write_Data }

PROCEDURE Func( B: Real;
      VAR Fb, Dfb: Real);

VAR
  I: Integer;
  S1, S2, S3, S4, S5, S6,
  Ex1, Ex2, Xi, X2, Yi, Y2: Real;
```

**Listing 10.3:** A least-squares fit to the nonlinearized exponential function (continued).

```
BEGIN
  S1 := 0.0;
  S2 := 0.0;
  S3 := 0.0;
  S4 := 0.0;
  S5 := 0.0;
  S6 := 0.0;
  FOR I:= 1 TO N DO
    BEGIN
      Xi := X[I];
      X2 := Xi * Xi;
      Yi := Y[I];
      Y2 := Yi * Yi;
      Ex1 := Exp(B * Xi);
      Ex[I] := Ex1;
      Ex2 := Ex1 * Ex1;
      S1 := S1 + Xi * Ex2/Y2;
      S2 := S2 + Ex1/Yi;
      S3 := S3 + Xi * Ex1/Yi;
      S4 := S4 + Ex2/Y2;
      S5 := S5 + 2.0 * X2 * Ex2/Y2;
      S6 := S6 + X2 * Ex1/Yi
    END;
  Fb := S1*S2 - S3*S4;
  Dfb := S2*S5 - S1*S3 - S4*S6;
  A := S2/S4
END; { Func }

PROCEDURE Newton(VAR X: Real);

CONST
  Tol = 1.0E-6;
  Max = 20;

VAR
  I: Integer;
  Fx, Dfx, Dx, X1: Real;

BEGIN  { Newton }
  Error := False;
  I := 0;
  REPEAT
    I := I + 1;
    X1 := X;
    Func(X, Fx, Dfx);
    IF Dfx = 0.0 THEN
      BEGIN
        Error := True;
        X := 1.0;
        WriteLn('ERROR: slope zero')
      END
```

**Listing 10.3:** A least-squares fit to the nonlinearized exponential function (continued).

```
             ELSE
               BEGIN
                 Dx := Fx / Dfx;
                 X  := X1 - Dx;
               END
          UNTIL
            Error  OR
            (I > Max) OR
            (Abs(Dx) <= Abs(Tol * X));
          IF I > Max THEN
            BEGIN
              WriteLn ('ERROR: no convergence in ',
                 Max, ' loops');
              Error := True
            END
      END; { Newton }

      PROCEDURE Nlin(X, Y: Ary;
                 VAR Y_Calc: Ary;
                       N: Integer);
      { fit the diffusion equation through
        N sets of X and Y pairs of points }

      VAR
        Resid: Ary;
        I: Integer;
        Xi, Yi, Sum_X,
        Sum_Y, Sum_Y2, B1,
        Sum_Xy, Sum_X2: Real;

      BEGIN { Nlin }
        Ncol := 2; { two terms }
        Sum_X := 0.0;
        Sum_Y := 0.0;
        Sum_Xy := 0.0;
        Sum_X2 := 0.0;
        FOR I := 1 TO N DO
          BEGIN
            Xi := X[I];
            Yi := Ln(Y[I]);
            Sum_X := Sum_X + Xi;
            Sum_Y := Sum_Y + Yi;
            Sum_Y2 := Sum_Y2 + Yi*Yi;
            Sum_Xy := Sum_Xy + Xi*Yi;
            Sum_X2 := Sum_X2 + Xi*Xi
          END;
        B := (Sum_Xy - Sum_X*Sum_Y/N) /
             (Sum_X2 - Sqr(Sum_X)/N);
        Newton(B);
        Coef[1] := A;
        Coef[2] := B;
        SRS := 0.0;
        FOR I := 1 TO N DO
```

**Listing 10.3:** A least-squares fit to the nonlinearized exponential function (continued).

```
      BEGIN
        Y_Calc[I] := A * Ex[I];
        IF Y[I] <> 0.0 THEN
          Resid[I] := Y_Calc[I] / Y[I] - 1.0
        ELSE
          Resid[I] := Y[I] / Y_Calc[I] - 1.0;
        SRS := SRS + Sqr(Resid[I])
      END
END; { Nlin }

BEGIN   { main program }
  Get_Data (X, Y, N);
  Nlin(X, Y, Y_Calc, N);
  Write_Data
END.
```

**Listing 10.3:** A least-squares fit to the nonlinearized exponential function (continued).

## *Running the Nonlinearized Exponential Curve-Fitting Program*

Create a file named NLIN3.PAS, and type the program shown in Listing 10.3. Run this program, and check that the results look like Figure 10.3.

Now compare the results shown in Figure 10.2 to those in Figure 10.3. The diffusion constant $D_0$ and the activation energy Q are about the same but the SRS is smaller for the nonlinearized fit. However, if we used the linearized residuals

$$r = \ln D_0 - \frac{Q}{RT}$$

in the SRS, the linearized SRS would be smaller. Furthermore, both linearized and nonlinearized approaches give the same results if the data precisely follow an exponential equation. Therefore, you should calculate both the linearized and nonlinearized curve fit. If the results are very different, then you should suspect an error in measuring or recording the data.

```
     I    T C        D              D Calc
     1   600  1.40000E-012   1.31052E-012
     2   650  5.50000E-012   5.41377E-012
     3   700  1.80000E-011   1.93305E-011
     4   750  6.10000E-011   6.09470E-011
     5   800  1.60000E-010   1.72657E-010
     6   850  4.40000E-010   4.45807E-010
     7   900  1.20000E-009   1.06167E-009

     Coefficients
     3.08929E-001      Constant term
     -2.2860E+004

     D0 =      0.31 cm sq/sec.
     Q =      45.42 kcal/mole

     SRS =     0.0295
```

**Figure 10.3:** The diffusion of zinc in copper (nonlinear fit).

# Summary

We have seen examples of two approaches to nonlinear curve fitting. The forms of the equations we used in our examples were (1) the rational function, and (2) the exponential function. We used both a linearization and a direct approach. It is important to remember that neither of the approaches we have studied represents a general method for nonlinear curve fitting; rather we have examined particular techniques that can be used to solve specific curve-fitting equations.

# Exercises

**10-1.** The band-gap energy, $E_g$, of an intrinsic semiconductor can be determined from the expression

$$1/\rho = \sigma = Ae^{-(E_g/2kT)}$$

where $\rho$ is the electrical resistivity, $\sigma$ is the electrical conductivity, k is the Boltzmann constant, and T is the temperature. The following experimental data were obtained from an intrinsic thermistor:

| T (°C) | R (ohms) |
|--------|----------|
| 0      | 1380     |
| 4      | 1200     |

| | |
|---|---|
| 10 | 880 |
| 18 | 660 |
| 25 | 488 |
| 38 | 304 |
| 43 | 248 |
| 54 | 170 |
| 62 | 139 |
| 77 | 83 |

Linearize the equation by taking the logarithm, and then perform a least-squares fit using the experimental data. Since we are only interested in the slope of the resulting fit, we do not have to convert resistance to resistivity. (The temperature data must be converted to kelvins.) Use a Boltzmann's constant of $8.61 \times 10^{-5}$ eV/K to obtain the band-gap energy in electron volts.

**10-2.** The time-temperature relationship for the crystallization of a material follows the expression:

$$\frac{1}{t} = Ae^{-Q/RT}$$

where t is the time, Q is the activation energy, R is the gas constant, and T is the temperature. Find the activation energy Q and coefficient A from the given data by performing a least-squares fit on the linearized form of the equation:

$$\ln t = -\ln A + Q/RT$$

Be sure to convert the temperature to kelvins. For the gas constant, use R = 1.987 cal/deg mole.

| T (°C) | t (min) |
|---|---|
| 350 | 49 |
| 360 | 40 |
| 370 | 34 |
| 380 | 28 |
| 390 | 24 |
| 400 | 20 |
| 410 | 17 |
| 420 | 14 |
| 430 | 12 |

# Advanced
# Applications

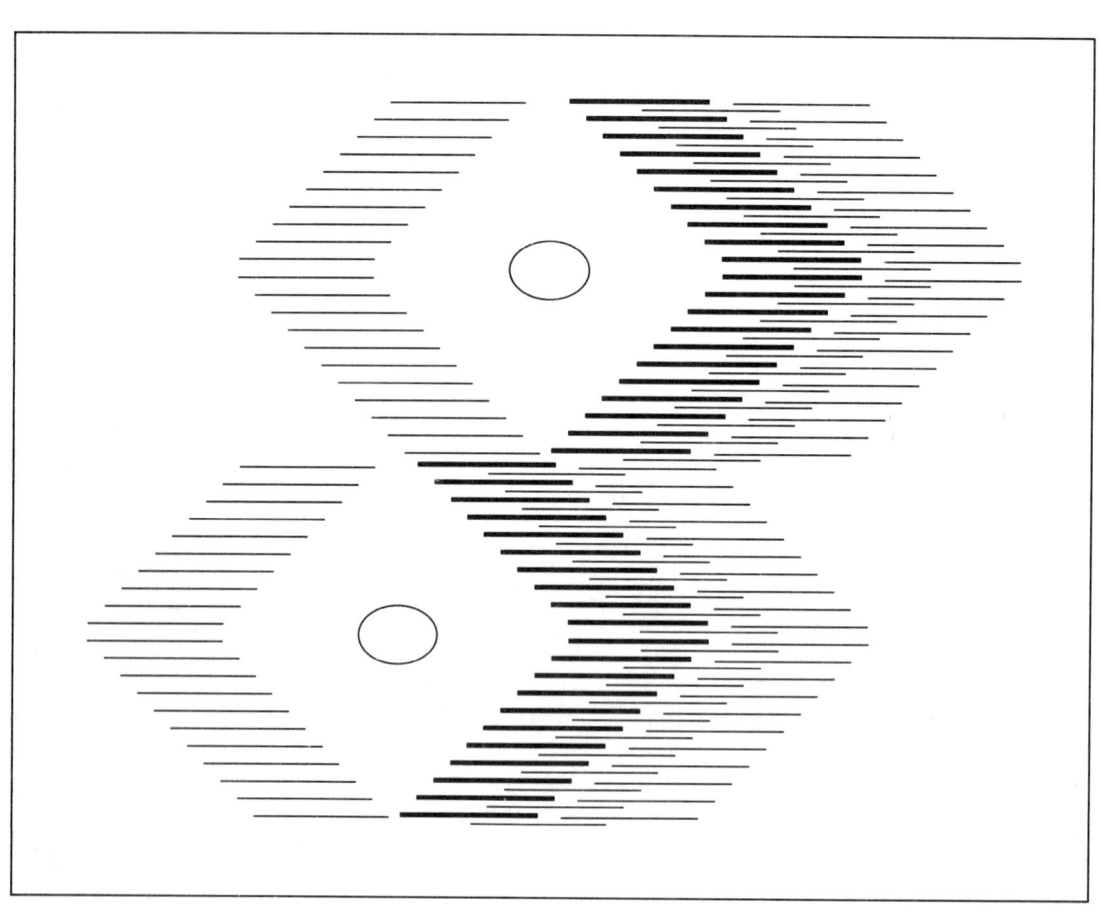

Chapter *11*

In this chapter, we will consider several advanced mathematical applications. We will study the Gaussian error function, the Gamma function, and the Bessel functions.

For evaluating the Gaussian error function, we will write a program that uses Simpson's method for numerical integration and a program that uses an infinite series expansion.

We then will study the special properties of the Gamma function and develop a program to evaluate it. We will also use this program to investigate numerical solutions to the Bessel equation. We will consider Bessel functions of the first and second kind.

Let us begin by reviewing the distribution function.

# The Normal and Cumulative Distribution Functions

As we discussed in Chapter 2, random errors, introduced during experimental measurement, cause a sequence of observed values to be dispersed about the mean, or average. A frequency plot of the resulting data shows a bell-shaped curve. This shape, shown in Figure 11.1, is described as a normal distribution, or probability density, function.

The normal distribution is defined by the equation

$$f(x) = \frac{e^{\frac{-x^2}{2}}}{\sqrt{2\pi}} \qquad (1)$$

This function has a peak, or mean value, at x = 0, and ranges from minus infinity to plus infinity. The area under the probability-density curve (above the x axis) is normalized to a value of unity. That is

$$\int_{-\infty}^{\infty} f(x)dx = 1 \qquad (2)$$

From the symmetry of the curve, it can be seen that the area from x = 0 to infinity (the right half of the curve) is equal to one-half the total area.

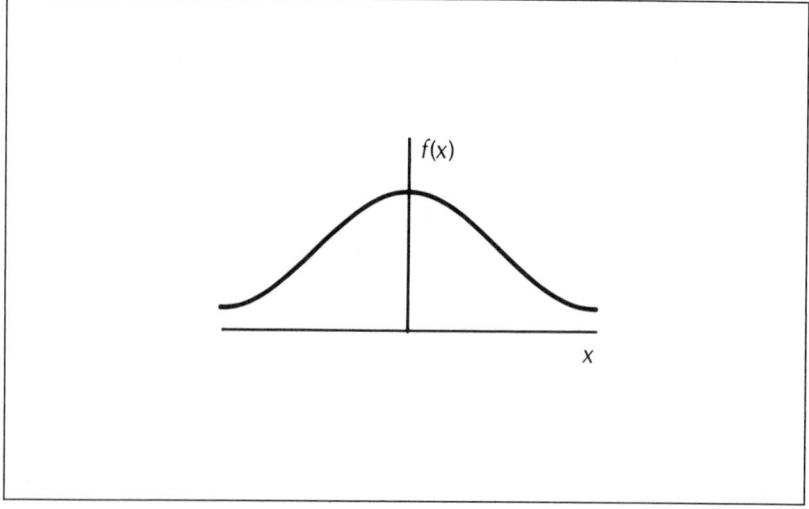

**Figure 11.1:** The normal distribution function.

The area from x $= -\infty$ to x $=$ b is called the *cumulative distribution function* F(x):

$$F(x) = \int_{-\infty}^{b} f(x)dx \tag{3}$$

This integral cannot be solved in closed form, but it is tabulated in the handbook *CRC Standard Mathematical Tables* by W. Beyer (CRC Press, West Palm Beach, Florida), which is published annually. Sometimes the integral

$$G(x) = \int_{0}^{b} f(x)dx \tag{4}$$

is given instead. Because the curve is normalized, either integral can readily be obtained from the other. The relationship is

$$F(x) = G(x) + 0.5$$

In the next section, we will consider some numerical methods for determining the area G(x). However we will first finish our discussion of the normal curve.

## The Standard Deviation

The area under the normal curve of the measured values is related to the standard deviation. A small standard deviation corresponds to a close grouping of the measured values about the mean. Conversely, a large standard deviation corresponds to values that are spread further from the mean. Two normal distributions are shown in Figure 11.2. They both have the same mean value, but they have different standard deviations. The curve with the smaller standard deviation, or dispersion, has the higher peak at zero. However, both have the same area underneath the curve.

Figure 11.3 shows a plot of the distribution function described by Equation 3. This function has a value of 0.5 at x $= 0$ and rises asymptotically to a value of unity as x approaches infinity.

The function G(x) can be used to find the relationship between the standard deviation and the corresponding fraction of a particular sample. For example, G(1) has a value of 0.34. Thus, 34 percent of the

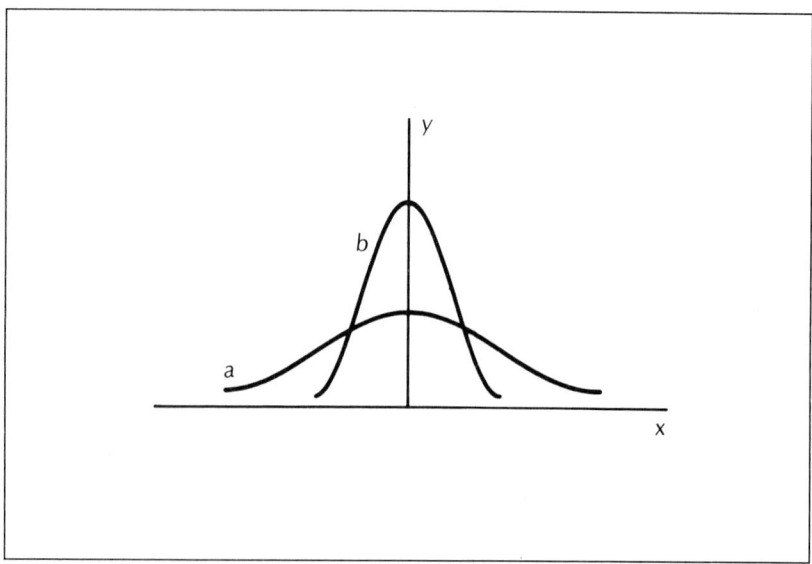

**Figure 11.2:** Normal curves with large (a) and small (b) standard deviations.

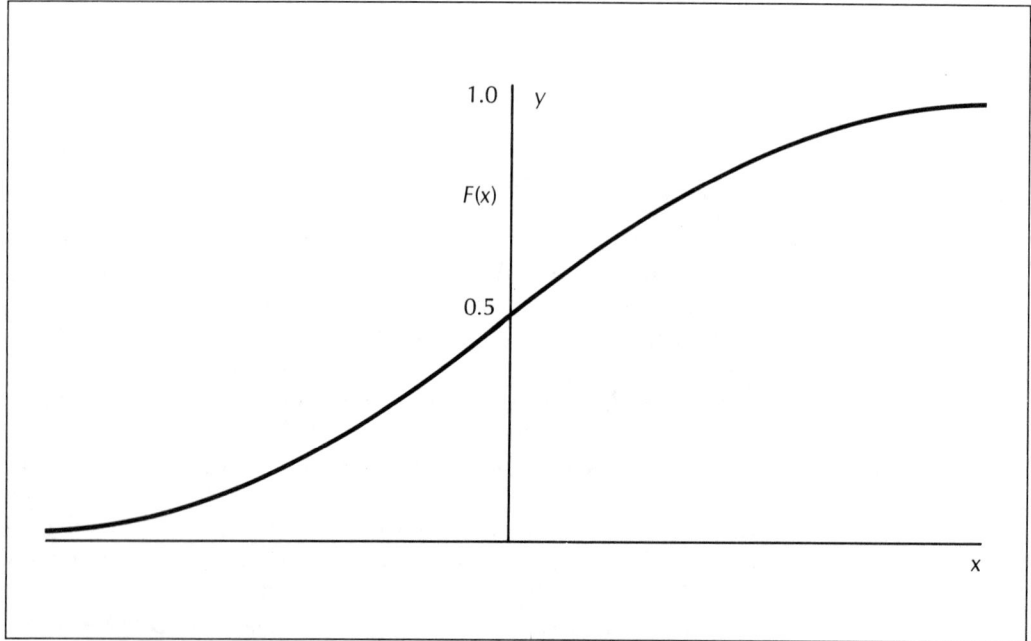

**Figure 11.3:** The cumulative distribution function.

population lies between a standard deviation of 0 and 1. Twice that value, 68 percent, corresponds to a range of 1 sigma on both sides of the mean. The distribution function can be readily obtained from the Gaussian error function, which we will discuss next.

# *The Gaussian Error Function*

Before formulating the Gaussian error function and writing a program to evaluate it, we will digress slightly to a common application area—*diffusion*. Our study of an equation that calculates the diffusion of one kind of atom into another will illustrate the usefulness of the Gaussian error function.

## *Diffusion in a One-Dimensional Slab*

*Diffusion* is the net flow of atoms, electrons, or heat from a more concentrated region to a less concentrated region. The resulting flux, J, can be described by *Fick's first law*:

$$J = -D\frac{dC}{dx}$$

where C is the concentration, and x is the distance. The *diffusion coefficient,* or *diffusivity,* D is the same quantity that we used in Chapter 10 to describe the diffusion of zinc in copper. The minus sign means that the flux occurs in a direction that is opposite to the concentration gradient. If we are concerned with the flow of atoms, J is expressed in atoms per unit area-second, and the concentration, C, is given in atoms per unit volume.

Fick's first law can also describe the flow of electrons in a conductor. We then write

$$J = \sigma\frac{dV}{dx}$$

Here J is the current density, V is the voltage, X is the distance, and $\sigma$ is the electrical conductivity.

In a similar way, the flow of heat can be expressed as

$$J = k\frac{dT}{dx}$$

where k is the thermal conductivity in units of energy/ area-second, and T is the temperature.

Fick's second law:

$$\frac{\partial C}{\partial t} = D \frac{\partial^2 C}{\partial x^2}$$

describes the concentration as a function of position and time. There are many solutions to Fick's second law, depending on the boundary conditions. For example, steady-state conditions occur when the concentration no longer changes with time. Since

$$\frac{\partial C}{\partial t} = 0$$

then

$$\frac{D \partial^2 C}{\partial x^2} = 0$$

This implies that a concentration gradient, $\partial C/\partial X$, is uncurving, or straight.

Another useful solution to Fick's second law describes the diffusion of one kind of atom into another. The surface concentration of the diffusing species is kept constant. The other species is in the shape of a one-dimensional, semi-infinite slab. For this arrangement, the solution to Fick's law is

$$\frac{C_x - C_0}{C_s - C_0} = 1 - \text{erf}(y) = 1 - \text{erf}\left(\frac{x}{2\sqrt{Dt}}\right)$$

Here, $C_x$ is the concentration of the diffusing species at time t and a distance x from the surface. $C_s$ is the surface concentration that is always constant, and $C_0$ is the initial, uniform concentration for all x at time equal to zero. D is the diffusion coefficient, and y has the value

$$\frac{x}{2\sqrt{Dt}}$$

If the initial concentration, $C_0$, is zero, then the equation reduces to

$$\frac{C_x}{C_s} = 1 - \text{erf}(y) = 1 - \text{erf}\left(\frac{x}{2\sqrt{Dt}}\right)$$

The quantity erf is the Gaussian error function. It is defined as

$$\text{erf}(y) = \frac{2}{\sqrt{\pi}} \int_0^y e^{-t^2} dt \qquad (5)$$

The functions F(x) (given in Equation 3) and G(x) (given in Equation 4) can be obtained from the error function by the relationship

$$F(x) = \frac{1 + \text{erf}\left(\frac{x}{\sqrt{2}}\right)}{2}$$

$$G(x) = \frac{\text{erf}\left(\frac{x}{\sqrt{2}}\right)}{2}$$

For example, the range represented by two standard deviations on either side of the mean can be found by the error function

$$2G(2) = \text{erf}\left(\frac{2}{\sqrt{2}}\right) = 95\%$$

# *Evaluating the Gaussian Error Function Using Simpson's Method*

The error function cannot be solved in closed form. However, we can obtain particular solutions. A straightforward approach is to use a numerical integration technique such as Simpson's method. Listing 11.1 gives a program for finding the error function in this way. Type this program, giving it the file name ERFSIMP.PAS.

## *Calculating the Error Function Using Simpson's Method*

Run the program. It repeatedly cycles, asking you for input. It calculates the error function by using Simpson's method, then displays the argument and the corresponding error function. You can terminate the program by entering a negative value or by pressing Ctrl-C. The program is adequate for solving functions with small arguments; however, its execution time increases and the accuracy decreases as the argument grows larger. Now we will consider another method that is not as limited.

```
PROGRAM Erfsimp;
{ integration by Simpson's method }

CONST
  Tol = 1.0E-5;

VAR
  Done: Boolean;
  Sum, Upper, Lower, Erf,
  Twopi: Real;

FUNCTION Fx( X: Real): Real;
BEGIN
  Fx := Exp(-X * X)
END;    { Function Fx }

PROCEDURE Simps(Lower, Upper, Tol: Real;
                            VAR Sum: Real);
{ numerical integration by Simpson's method
  function is Fx, limits are Lower and Upper
  with number of regions equal to Pieces.
  Partition is Delta_X, answer is Sum }

VAR
  I, Pieces: Integer;
  X, Delta_X, Even_Sum,
  Odd_Sum, End_Sum, Sum1: Real;

BEGIN
  Pieces := 2;
  Delta_X := (Upper - Lower) / Pieces;
  Odd_Sum := Fx(Lower + Delta_X);
  Even_Sum := 0.0;
  End_Sum := Fx(Lower) + Fx(Upper);
  Sum := (End_Sum + 4.0*Odd_Sum) * Delta_X / 3.0;
  REPEAT
    Pieces := Pieces * 2;
    Sum1 := Sum;
    Delta_X := (Upper - Lower) / Pieces;
    Even_Sum := Even_Sum + Odd_Sum;
    Odd_Sum := 0.0;
    FOR I:= 1 TO Pieces DIV 2 DO
      BEGIN
        X := Lower + Delta_X * (2.0*I -1.0);
        Odd_Sum := Odd_Sum + Fx(X)
      END;
    Sum := (End_Sum + 4.0*Odd_Sum
            + 2.0*Even_Sum) * Delta_X / 3.0
  UNTIL Abs(Sum1 - Sum) <= Abs(Tol * Sum1)
END; { Simps }
```

**Listing 11.1:** Calculating the Gaussian error function with Simpson's method.

```
BEGIN { main program }
  Done := False;
  Twopi := 2.0/Sqrt(Pi);
  Lower := 0.0;
  REPEAT
    WriteLn;
    Write(' Erf? ');
    ReadLn(Upper);
    IF Upper < 0.0 THEN
      Done := True
    ELSE IF Upper = 0.0 THEN
      WriteLn(' Erf of 0.0 is 0.0')
    ELSE     { Upper > 0 }
      BEGIN
        Simps(Lower, Upper, Tol, Sum);
        Erf := Twopi * Sum;
        WriteLn(' Erf of ', Upper:7:2,
          ', is ', Erf:8:4)
      END
  UNTIL Done
END.
```

**Listing 11.1:** Calculating the Gaussian error function with Simpson's method (continued).

## *Evaluating the Gaussian Error Function Using an Infinite Series Expansion*

Another way to evaluate the error function is to substitute an infinite series. The new expression is then integrated term by term to produce another infinite series. The result is

$$\text{erf}(y) = \frac{2}{\sqrt{\pi}} e^{-y^2} \sum_{n=0}^{\infty} \frac{2^n y^{2n+1}}{1 \cdot 3 \cdot \ldots \cdot (2n + 1)}$$

Listing 11.2 gives a program for calculating the error function in this way. Type the program, using the file name ERF.PAS. When you run the program, it asks you to enter an argument to the error function, then it displays the argument and the corresponding function.

The infinite series is evaluated in procedure Erf. The procedure adds each new term to the sum of the previous term. If a particular term does not change the sum by more than the value of Tol, then the procedure terminates the routine and displays the current value.

```
PROGRAM Erfd;
{ evaluation of the Gaussian error function }

VAR
  X, Ans, SqrtPi: Real;
  Done: Boolean;

FUNCTION Erf(X: Real): Real;

{ Infinite series expansion of the
  Gaussian error function }

CONST
  Tol = 1.0E-6;

VAR
  X2, Sum, Sum1, Term: Real;
  I: Integer;
BEGIN
  IF X = 0.0 THEN Erf := 0.0
  ELSE IF X > 4.0 THEN Erf := 1.0
  ELSE
    BEGIN
      X2  := X*X;
      Sum := X;
      Term := X;
      I := 0;
      REPEAT
        I := I+1;
        Sum1 := Sum;
        Term := 2.0 * Term * X2/(1.0 + 2.0*I);
        Sum  := Term + Sum1
      UNTIL Term < Tol * Sum;
      Erf := 2.0 * Sum * Exp(-X2) / SqrtPi
    END  { IF }
END; { Erf }

BEGIN  { main program }
  SqrtPi := Sqrt(Pi);
  Done := False;
  WriteLn;
  REPEAT
    Write(' Arg? ');
    ReadLn( X);
    IF X < 0.0 THEN
      Done := True
    ELSE
      BEGIN
        Ans := Erf(X);
        WriteLn('Erf of ', X:6:3,
          ' is ', Ans:9:5)
      END
  UNTIL Done
END.
```

Listing 11.2: An infinite series expansion for the Gaussian error function.

### *Running the Infinite Series Expansion Program*

Run this version, entering values from 0 to 2. The resulting function has a range from zero to unity. The result is 0 if the argument is 0, and it approaches 1 for arguments above 3. On the other hand, the function is approximately equal to its argument for the range 0 to 0.6. The error function has the same relative shape as the cumulative distribution function shown in Figure 11.3. Selected values of the error function are given in Figure 11.4. The program will cycle, so you can find values for different error functions. You can terminate the program by pressing Ctrl-C. Now let us see how to use this program.

### *Using the Infinite Series Expansion Program*

Suppose that zinc is to be diffused into a bar of copper at 900°C. We can use the error function to determine the concentration of copper as a function of time and the distance from the surface. We can calculate the diffusion coefficient by using the diffusion constant and activation energy values that we obtained in Chapter 10 with the nonlinear curve-fitting program.

$$D = 0.31 \, e^{\frac{-45,420}{1.987(900+273)}}$$

| y | erf(y) |
|-----|--------|
| 0.0 | 0.0 |
| 0.1 | 0.1125 |
| 0.2 | 0.2227 |
| 0.3 | 0.3286 |
| 0.4 | 0.4284 |
| 0.5 | 0.5205 |
| 0.7 | 0.6778 |
| 1.0 | 0.8427 |
| 2.0 | 0.9953 |

**Figure 11.4:** Selected values of the Gaussian error function.

If the distance is chosen to be 0.01 cm and the time is taken as 13 hours, then the argument y in the error function becomes 0.7. This corresponds to an error function of 0.67 and an error function complement of 0.33, which means that the concentration of zinc at a depth of 0.01 cm is 33 percent of the surface concentration after 13 hours.

In the next section, we will see that roundoff errors make it difficult to find a direct solution to the error function complement. We will examine a program that uses different functions for small and large arguments.

# The Complement of the Error Function

For the solution to the diffusion equation for zinc into copper, we need the complement of the error function rather than the error function itself. The complement is defined as

$$\text{erfc}(y) = \frac{2}{\sqrt{\pi}} \int_{y}^{\infty} e^{-t^2} dt \qquad (6)$$

The complement is obtained from the relationship

$$\text{erfc}(y) = 1 - \text{erf}(y)$$

However, this calculation produces large roundoff errors for arguments above 3. As erf approaches unity, $1 - \text{erf}$ approaches zero. Ultimately, all significant figures are lost. Furthermore, the computation time increases as the argument grows larger.

Alternatively, Equation 6 cannot be integrated by the trapezoidal method or by Simpson's method. A problem occurs in selecting the upper limit for the integral. A value larger than 8 will produce a floating-point underflow because $e^{-64}$ is so small. Yet the area from this point to infinity is significant and cannot be ignored.

## Evaluating the Complement of the Error Function

One solution to this problem is to provide two functions, one for small arguments and another for larger arguments. The program given in Listing 11.3 uses this approach. The infinite series expansion of erf, given in Listing 11.2, is used for smaller arguments. The procedure for larger arguments calculates the complementary error function from an asymptotic expansion. This algorithm becomes more accurate as the

```
PROGRAM Erfd3;
{ evaluate the Gaussian error function }

VAR
  X, Er, Ec, SqrtPi: Real;
  Done: Boolean;

FUNCTION Erf(X: Real): Real;
{ Infinite series expansion of the
  Gaussian error function }

CONST
  Tol = 1.0E-6;

VAR
  X2, Sum, Sum1, Term: Real;
  I: Integer;

BEGIN
  X2 := X*X;
  Sum := X;
  Term := X;
  I := 0;
  REPEAT
    I := I+1;
    Sum1 := Sum;
    Term := 2.0 * Term * X2/(1.0 + 2.0*I);
    Sum := Term + Sum1
  UNTIL Term < Tol * Sum;
  Erf := 2.0 * Sum * Exp(-X2) / SqrtPi
END; { Erf }

FUNCTION Erfc(X: Real): Real;
{ Complement of error function }

CONST
  SqrtPi = 1.7724538;
  Terms = 12;

VAR
  X2, U, V, Sum: Real;
  I: Integer;

BEGIN
  X2 := X*X;
  V := 1.0 / (2.0*X2);
  U := 1.0 + V*(Terms+1.0);
  FOR I := Terms DOWNTO 1 DO
    BEGIN
      Sum := 1.0 + I*V/U;
      U := Sum
    END;
  Erfc :=  Exp(-X2) / (X * Sum * SqrtPi)
END; { Erfc }
```

**Listing 11.3:** The error function and its complement.

```
BEGIN { main program }
  SqrtPi := Sqrt(Pi);
  Done := False;
  WriteLn;
  REPEAT
    Write(' Arg? ');
    ReadLn( X);
    IF X < 0.0 THEN
      Done := True
    ELSE
      BEGIN
        IF X = 0.0 THEN
          BEGIN
            Er := 0.0;
            Ec := 1.0
          END
        ELSE
          BEGIN
            IF X < 1.5 THEN
              BEGIN
                Er := Erf(X);
                Ec := 1.0 - Er
              END
            ELSE
              BEGIN
                Ec := Erfc(X);
                Er := 1.0 - Ec
              END   { IF }
          END;
        WriteLn(' X= ', X:6:2, ', Erf= ',
          Er:7:4, ', Erfc= ', Ec:10)
      END   { IF }
  UNTIL Done
END.
```

**Listing 11.3:** The error function and its complement (continued).

argument increases. The equation, expressed as a continued fraction rather than the usual infinite series, is

$$\text{erfc}(y) = \frac{1/[1 + v/\{1 + 2v/[1 + 3v/(1 + \ldots)]\}]}{\sqrt{\pi}\, y e^{y^2}}$$

$$v = \frac{1}{2y^2}$$

### *Running the Error Function Complement Program*

Copy Listing 11.2, giving the new version the name ERFC.PAS, and change it to look like Listing 11.3. When you run the new version, it will display both the error function and its complement. For arguments that are less than 1.5, the program calculates the error function in procedure Erf using the infinite series expansion given in Listing 11.2. The complement is then obtained by subtraction from unity. On the other hand, if the argument is 1.5 or larger, the program calculates the complement from the asymptotic expansion using procedure Erfc. The error function is determined by subtraction from unity.

Compare the error function from this version with the data given in Figure 11.4. Then try the values given in Figure 11.5 to check the complementary error function.

In the next section, we will examine the special properties of the Gamma function, and we will write a program to evaluate it.

| y | erfc(y) |
|-----|-----------|
| 1.5 | 3.390E-2 |
| 2.0 | 4.678E-3 |
| 2.5 | 4.070E-4 |
| 3.0 | 2.209E-5 |
| 3.5 | 7.431E-7 |
| 4.0 | 1.542E-8 |
| 4.5 | 1.966E-10 |

**Figure 11.5:** The complementary error function.

## *The Gamma Function*

The Gamma function is related to the error function; it is defined by the integral

$$\Gamma(n) = \int_0^\infty x^{n-1} e^{-x} dx \tag{7}$$

The Gamma function is important because it is part of the solution to Bessel's equation. (Bessel's equation is used for studying the vibration of circular membranes and the conduction of heat and electricity flowing in cylindrical conductors.) In addition, the Gamma function can be used to calculate factorials because of the recursive relationship

$$\Gamma(n + 1) = n\Gamma(n) \tag{8}$$

Since $\Gamma(1) = 1$, we can see that

$$\Gamma(2) = \quad \Gamma(1) = 1 \qquad = 1!$$
$$\Gamma(3) = 2\Gamma(2) = 1 \cdot 2 \qquad = 2!$$
$$\Gamma(4) = 3\Gamma(3) = 1 \cdot 2 \cdot 3 = 3!$$

$$\cdots$$

$$\Gamma(n) = (n - 1)\Gamma(n - 1) = (n - 1)!$$

Thus, the general formula for using the Gamma function to calculate factorials is

$$\Gamma(n + 1) = n\Gamma(n) = n!$$

Since the Gamma function is defined for all real arguments greater than zero, the factorial of noninteger arguments can be defined as well. In addition, we find that

$$\Gamma(0.5) = -0.5! = \sqrt{\pi}$$

This relationship is useful in calculating Bessel functions. The Gamma function is also defined for noninteger negative numbers. Its value, however, is infinite for zero and for negative integers. A plot of the Gamma function for real arguments is given in Figure 11.6.

The Gamma function can be calculated using a form of Stirling's approximation:

$$\Gamma(x) = \sqrt{\frac{2\pi}{x}}\, x^x e^y$$

$$y = \frac{1}{12x} - \frac{1}{360x^3} - x$$

Since this is an asymptotic expression, the accuracy of the calculation improves as the argument increases.

### *Evaluating the Gamma Function*

The program given in Listing 11.4 uses Stirling's approximation to calculate the Gamma function. Arguments can be any real number greater than 0 and less than about 32 for the regular Turbo Pascal compiler; with Turbo-87, you can enter a number as large as 169. Arguments can also be negative if they are nonintegral. However, the argument cannot be zero or a negative integer.

The program increments positive arguments by 2 and calculates the Gamma function of the new argument. It then reduces the resulting

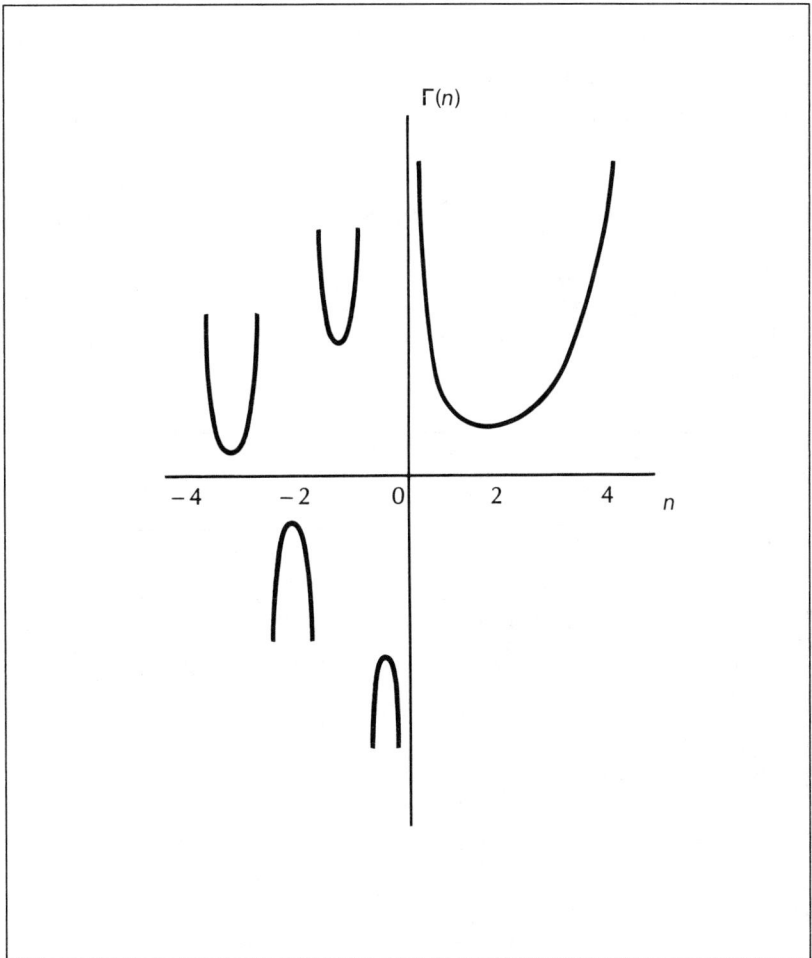

**Figure 11.6:** The Gamma function.

value to the corresponding original argument by using the algorithm

$$\Gamma(x) = \frac{\Gamma(x + 2)}{x(x + 1)}$$

This conversion is not necessary for larger arguments, but it ensures that there will be at least six figures of precision for all values of x.

Negative arguments are incremented until they are positive, and the Gamma function is called recursively to calculate the new value. The result is then corrected for the original argument.

## *Running the Gamma Function Testing Program*

Create a file named GAMMA.PAS, and type the program shown in Listing 11.4. Run the program, and enter the values given in Figure 11.7 to be sure that you get the same results. Enter a number smaller than $-22$ or press Ctrl-C to terminate the program. Since x! $= \Gamma(x + 1)$, you can rewrite the program so that it generates factorials rather than the Gamma function.

The final two sections of this chapter introduce the Bessel functions of the first and second kind. These functions can be used to solve Bessel's equation, a differential equation with many mathematical and scientific applications.

```
PROGRAM Tstgam;
  { test the Gamma function }

VAR
  X: Real;

FUNCTION Gamma(X: Real): Real;

VAR
  I, J: Integer;
  Y, Gam: Real;

BEGIN     { Gamma function }
  IF X >= 0.0 THEN
    BEGIN
      Y := X + 2.0;
      Gam := Sqrt(2 * Pi/Y)
        * Exp(Y*Ln(Y) + (1 - 1/(30*Y*Y))/(12*Y)-Y);
      Gamma := Gam / (X * (X+1))
    END
```

Listing 11.4: Calculation of the Gamma function.

```
ELSE   { X < 0 }
  BEGIN
    J := 0;
    Y := X;
    REPEAT { increment argument until positive }
      J := J + 1;
      Y := Y + 1.0
    UNTIL Y > 0.0;
    Gam := Gamma(Y); { recursive call }
    FOR I := 0 TO J-1 DO
      Gam := Gam / (X + I);
    Gamma := Gam
  END   { if }
END; { Gamma function }

BEGIN  { main program }
  WriteLn;
  REPEAT
    REPEAT
      Write(' X: ');
      ReadLn( X)
    UNTIL X <> 0.0;
    WriteLn(' Gamma is ', Gamma(X):9:4)
  UNTIL X < -22.0
END.
```

**Listing 11.4:** Calculation of the Gamma function (continued).

| x | $\Gamma(x)$ | |
|---|---|---|
| 1 | 1 | 0! |
| 2 | 1 | 1! |
| 3 | 2 | 2! |
| 4 | 6 | 3! |
| 5 | 24 | 4! |
| 6 | 120 | 5! |
| 0.5 | 1.7725 | $\sqrt{\pi}$ |
| −0.5 | −3.5449 | $-\Gamma(0.5)/0.5$ |
| −1.5 | 2.3633 | $-\Gamma(-0.5)/1.5$ |

**Figure 11.7:** Selected values of the Gamma function.

# *Bessel Functions*

Bessel's equation

$$x^2y'' + xy' + (x^2 - n^2)y = 0$$

arises in the analysis of many different kinds of problems involving circular symmetry, such as studies of drum vibrations or heat conduction in cylinders. In this equation, x is the independent variable, y is the dependent variable, and N is a constant known as the order. This is a nonlinear differential equation that cannot be solved in closed form. One solution to Bessel's equation is:

$$y = J_n(x)$$

where J is the *n*-order Bessel function of the first kind.

The Bessel functions have been extensively tabulated for particular values of x and N. However, they are difficult to use in this form. For values of x less than about 15, the J Bessel functions can be calculated from the infinite series

$$J_n(x) = \sum_{k=0}^{n} \frac{(-1)^k}{k!\Gamma(n + k + 1)} \left(\frac{x}{2}\right)^{n+2k} \tag{9}$$

On the other hand, the asymptotic expression

$$J_n(x) = \sqrt{\frac{2}{\pi x}} \cos\left(x - \frac{\pi}{4} - \frac{n\pi}{2}\right) \tag{10}$$

can be used for calculating larger values of x. Selected values of the Bessel functions $J_0$ and $J_1$ are shown in Figure 11.8.

## *Programming Bessel Functions of the First Kind*

The program shown in Listing 11.5 uses both Equations 9 and 10 to calculate the Bessel functions of the first kind. It incorporates the Gamma function from the previous section. The order can be zero or a positive number. The argument can also be a noninteger negative number. The program uses the infinite series for arguments less than 15, and the asymptotic expression for larger arguments. Its calculations are the least precise for arguments between 13 to 17.

Type the program, give it the name BESSJ.PAS, and run it. Try the values shown in Figure 11.9 for the given order and argument, and compare the results to those shown in that figure. You can terminate the program by entering an order less than $-25$ or by pressing Ctrl-C.

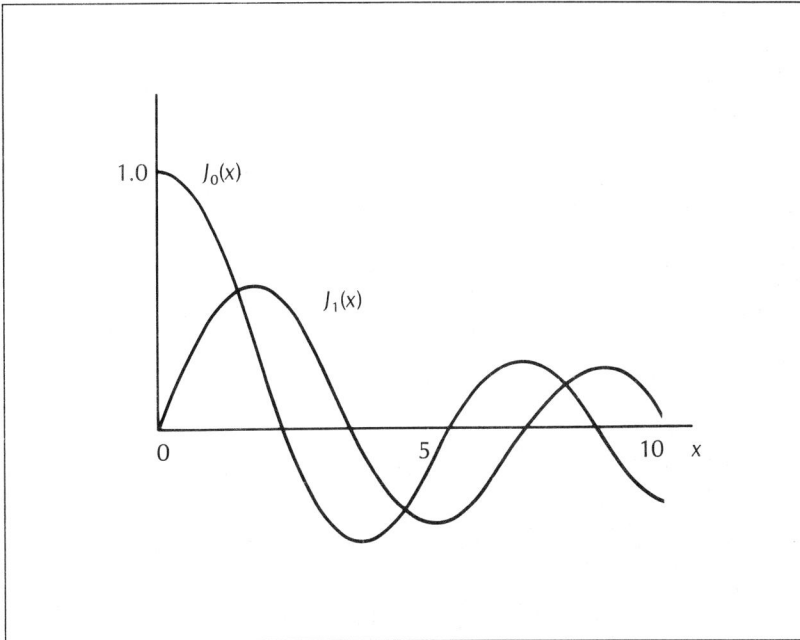

**Figure 11.8:** The Bessel functions $J_0$ and $J_1$.

```
PROGRAM Tstbes;
  { test Bessel function }
  { the Gamma function is required }

VAR
  Done: Boolean;
  X, Ordr: Real;

FUNCTION Gamma(X: Real): Real;

VAR
  I, J: Integer;
  Y, Gam: Real;

BEGIN    { Gamma function }
  IF X >= 0.0 THEN
    BEGIN
      Y := X + 2;
      Gam := Sqrt(2 * Pi/Y) * Exp(Y * Ln(Y)
          + (1 - 1/(30*Y*Y))/(12*Y)-Y);
      Gamma := Gam / (X * (X+1))
    END
```

**Listing 11.5:** Bessel functions of the first kind.

```
     ELSE  { X < 0 }
       BEGIN
         J := 0;
         Y := X;
         REPEAT
           J := J + 1;
           Y := Y + 1.0
         UNTIL Y > 0.0;
         Gam := Gamma(Y);
         FOR I := 0 TO J-1 DO
           Gam := Gam / (X + I);
         Gamma := Gam
       END    { if }
  END; { Gamma function }

  FUNCTION Bessj(X, N: Real): Real;
  { cylindrical Bessel function }
  { of the first kind }
  { the Gamma function is required }

  CONST
    Tol = 1.0E-5;

  VAR
    I: Integer;
    Term, New_Term, Sum, X2: Real;

  BEGIN { Bessj }
    X2 := X * X;
    IF (X = 0.0) AND (N = 1.0) THEN
      Bessj := 0.0
    ELSE IF X > 15 THEN { asymptotic expansion }
      Bessj := Sqrt(2/(Pi*X))*Cos(X - Pi/4 - N*Pi/2)
    ELSE
      BEGIN { regular infinite series }
        IF N = 0.0 THEN
          Sum := 1.0
        ELSE
          Sum := Exp(N * Ln(X / 2))/ Gamma(N+1.0);
        New_Term := Sum;
        I := 0;
        REPEAT
          I := I + 1;
          Term := New_Term;
          New_Term := -Term * X2 * 0.25/(I * (N + I));
          Sum := Sum + New_Term
        UNTIL Abs(New_Term) <= Abs(Sum * Tol);
        Bessj := Sum
      END  { if }
  END;  { Bessj }
```

**Listing 11.5:** Bessel functions of the first kind (continued).

```
BEGIN  { main program }
  Done := False;
  REPEAT
    Write(' Order: ');
    ReadLn( Ordr);
    IF Ordr < -25.0 THEN
      Done := True
    ELSE
      BEGIN
        Write(' X: ');
        ReadLn(X);
        WriteLn(' J Bessel is ', Bessj(X, Ordr):10:5)
      END
  UNTIL Done
END.
```

**Listing 11.5:** Bessel functions of the first kind (continued).

| Order | Argument | Function |
|-------|----------|----------|
| 1     | 1        | 0.4401   |
| 0     | 1        | 0.7652   |
| 1     | 0.5      | 0.2423   |
| 0     | 0.5      | 0.9385   |
| 1     | 10       | 0.04347  |
| 0     | 10       | $-0.2459$ |
| 0.25  | 1        | 0.7522   |
| 0.25  | 1.5      | 0.6192   |
| 0.5   | 1.5708   | 0.6366 $(2/\pi)$ |
| $-0.25$ | 1      | 0.6694   |
| $-0.25$ | 1.5    | 0.3180   |
| $-0.75$ | 1.5    | $-0.2684$ |

**Figure 11.9:** Selected Bessel function values.

Since Bessel's equation is second order, we need two independent solutions. When the order N is not an integer, both solutions can be obtained from the J Bessel functions. The formula is

$$y = AJ_n(x) + BJ_{-n}(x)$$

where A and B are constants to be determined from the boundary conditions.

## Programming Bessel Functions of the Second Kind

If the order of Bessel's equation is an integer, the solutions $J_n(x)$ and $J_{-n}(x)$ are linearly dependent. Then, a second, independent solution can be obtained from the formula

$$y = AJ_n(x) + BY_n(x)$$

where Y is a Bessel function of the second kind.

The program given in Listing 11.6 can be used to calculate Bessel functions of the second kind. While the order can be any real number, it is customary to use these functions only for integer orders. The program uses two different algorithms. For arguments less than 12, it calculates the values of $Y_0$ and $Y_1$ from the expression

$$Y_0(x) = \frac{2}{\pi} \sum_{m=0}^{n} (-1)^m \left(\frac{x}{2}\right)^{2m} \frac{\left[\ln \frac{x}{2} + \gamma - h\right]}{(m!)^2}$$

$$Y_1(x) = -\frac{2}{\pi x} + \frac{2}{\pi} \sum_{m=1}^{n} (-1)^{m+1} \left(\frac{x}{2}\right)^{2m-1} \frac{\left[\ln \left(\frac{x}{2}\right) + \gamma - h + \frac{1}{2m}\right]}{(m!)(m-1)!}$$

$$h = \sum_{r=1}^{m} \frac{1}{r} \quad \text{if} \quad m \geq 1$$

and $\gamma$ is Euler's constant ($\gamma = 0.57721566$). Orders other than 0 and 1 are calculated from the formula

$$Y_n(x) = \frac{2n}{x} Y_{n-1}(x) - Y_{n-2}(x)$$

For larger arguments, an asymptotic expansion similar to the one used for the J Bessel functions is employed:

$$Y_n(x) = \sqrt{\frac{2}{\pi x}} \sin \left(x - \frac{\pi}{4} - \frac{n\pi}{2}\right)$$

```
PROGRAM Besy;
{ evaluation of Bessel's function }
{ of the second kind }

VAR
  X, Ordr, Pi2: Real;
  Done: Boolean;

FUNCTION Bessy(X, N: Real): Real;
{ cylindrical Bessel function }
{ of the second kind }

CONST
  Small = 1.0E-8;
  Euler = 0.57721566;

VAR
  J: Integer;
  X2, Sum, Sum2, T, T2, Ts, Term, XX,
  Y0, Y1, Ya, Yb, Yc, Ans, A, B: Real;

BEGIN  { function Bessy }
  IF X < 12 THEN
    BEGIN
      XX := 0.5 * X;
      X2 := XX * XX;
      T := Ln(XX) + Euler;
      Sum := 0.0;
      Term := T;
      Y0 := T;
      J := 0;
      REPEAT
        J := J + 1;
        IF J <> 1 THEN Sum := Sum + 1/(J-1);
        Ts := T - Sum;
        Term := -X2*Term/(J*J) * (1 - 1/(J*Ts));
        Y0 := Y0 + Term
      UNTIL Abs(Term) < Small;
      Term := XX * (T - 0.5);
      Sum := 0.0;
      Y1 := Term;
      J := 1;
      REPEAT
        J := J + 1;
        Sum := Sum + 1/(J-1);
        Ts := T - Sum;
        Term := (-X2*Term) / (J*(J-1))*
                ((Ts - 0.5/J) / (Ts + 0.5/(J-1)));
        Y1 := Y1 + Term
      UNTIL Abs(Term) < Small;
      Y0 := Pi2 * Y0;
      Y1 := Pi2 * ( Y1 - 1/X);
      IF N = 0.0 THEN
        Ans := Y0
      ELSE IF N = 1.0 THEN
        Ans := Y1
```

**Listing 11.6:** Bessel functions of the second kind.

```
          ELSE
              BEGIN { find Y by recursion }
                  Ts := 2.0/X;
                  Ya := Y0;
                  Yb := Y1;
                  FOR J := 2 TO Trunc(N + 0.01) DO
                      BEGIN
                          Yc := Ts * (J-1) * Yb - Ya;
                          Ya := Yb;
                          Yb := Yc
                      END;
                  Ans := Yc
              END;
          Bessy := Ans;
      END  { X < 12 }
  ELSE  { X > 11, asymptotic expansion }
      Bessy := Sqrt(2/(Pi*X))*Sin(X - Pi/4 - N*Pi/2)
END; { function Bessy }

BEGIN { main program }
  Pi2 := 2.0/Pi;
  Done := False;
  WriteLn;
  REPEAT
      Write(' Order? ');
      ReadLn( Ordr);
      IF Ordr < 0.0 THEN
          Done := True
      ELSE
          BEGIN
              REPEAT
                  Write(' Arg? ');
                  ReadLn( X)
              UNTIL X >= 0.0;
              WriteLn(' Y Bessel is ', Bessy(X, Ordr):8:4)
          END  { IF }
  UNTIL Done
END.
```

**Listing 11.6:** Bessel functions of the second kind (continued).

A plot of the functions $Y_0$ and $Y_1$ is given in Figure 11.10.

Type the program, give it the file name BESSY.PAS, and run it. The argument must be a positive number, since the function goes to minus infinity at zero. Some typical values are shown in Figure 11.11. You can terminate the program by entering a negative number or by pressing Ctrl-C.

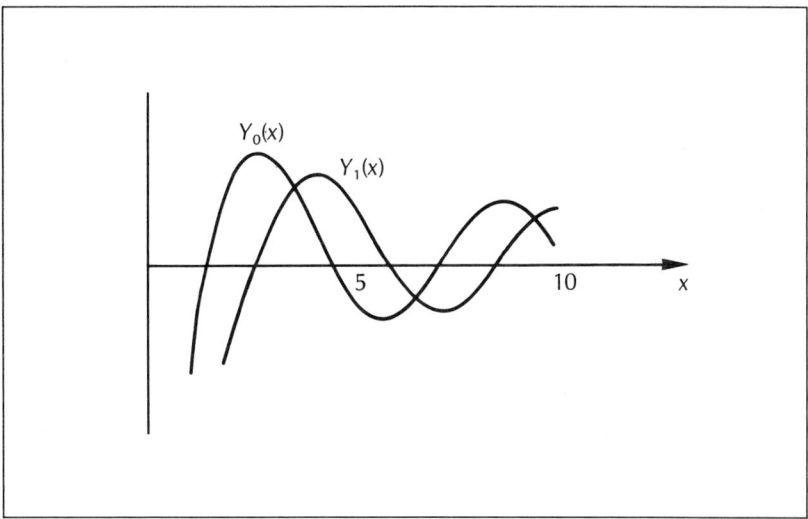

**Figure 11.10:** The Bessel functions $Y_0$ and $Y_1$.

| Order | Argument | Function |
|:-----:|:--------:|:--------:|
| 1 | 1 | −0.781 |
| 0 | 1 | 0.088 |
| 1 | 2 | −0.107 |
| 0 | 2 | 0.510 |
| 1 | 3 | 0.325 |
| 0 | 3 | 0.377 |

**Figure 11.11:** Selected values of the Bessel function of the second kind.

## *Summary*

In this chapter, we reviewed several concepts and developed some new tools that allow us to program advanced mathematical applications. We create Turbo Pascal programs that evaluate the Gaussian error function, the Gamma function, and the Bessel functions. These, and all the other programs presented in this book, demonstrate the effective use of Turbo Pascal for solving typical scientific and engineering problems.

## *Exercises*

**11-1.** Make a copy of the program shown in Listing 11.2, and alter the new version so that it calculates the distribution function $G(x)$ described by Equation 4. Show that $G(2) = 0.4772$.

**11-2.** Copy Listing 11.4 and alter the new version so that it calculates factorials. Show that the factorial of 0.5 is 0.886.

**11-3.** Write a program to calculate combinations. Start with the program used in the previous problem. The number of ways that six things can be taken two at a time, 6C2, is calculated from the expression

$$\frac{6!}{2! \, (6 - 2)!}$$

The program should request two numbers: the total number of items and the number of items taken at one time. Show that 6C2 has a value of 15.

# Reserved Words and Standard Identifiers

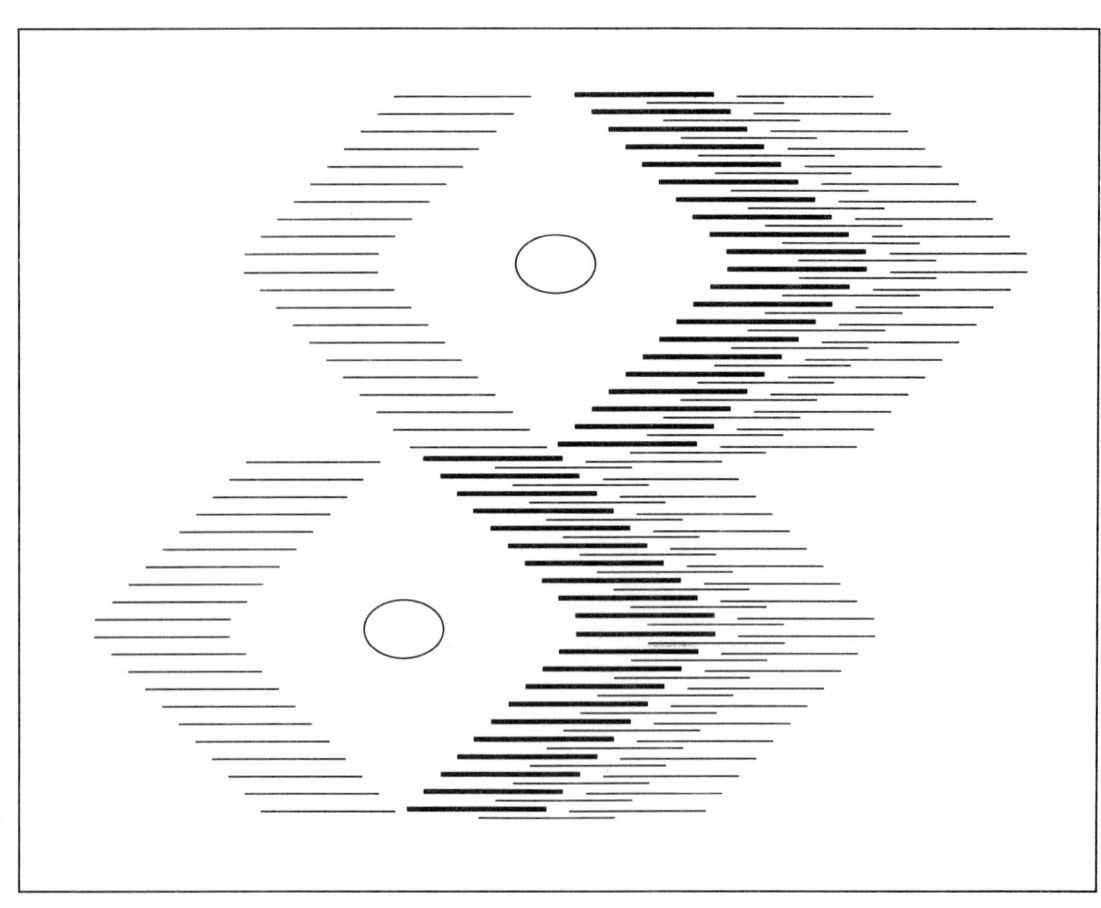

*Appendix* **A**

## Reserved Words

The following Turbo Pascal reserved words appear in boldface in this book. These words cannot be redefined.

| | | | | |
|---|---|---|---|---|
| **AND** | **ARRAY** | **BEGIN** | **CASE** | **CONST** |
| **DIV** | **DO** | **DOWNTO** | **ELSE** | **END** |
| **FILE** | **FOR** | **FUNCTION** | **GOTO** | **IF** |
| **IN** | **LABEL** | **MOD** | **NOT** | **OF** |
| **OR** | **PROCEDURE** | **PROGRAM** | **RECORD** | **REPEAT** |
| **SET** | **STRING** | **THEN** | **TO** | **TYPE** |
| **UNTIL** | **VAR** | **WHILE** | **WITH** | |

## *Standard Identifiers*

The following standard identifiers are defined in Turbo Pascal. However, they may be redefined.

| Name | Function |
|------|----------|
| Abs | Absolute value |
| Arctan | Arc tangent |
| Assign | Associate symbol to disk file |
| Boolean | Define logical variable |
| Char | Define ASCII variable |
| Chr | Convert integer to ASCII character |
| Close | Release file handle when done with file |
| Cos | Cosine |
| Eof | True when end of file is reached |
| Exit | Leave block |
| Exp | e raised to power |
| False | Value of Boolean variable |
| Integer | Define integer variable |
| KeyPressed | True when key pressed |
| Odd | True if integer argument is odd |
| Pi | The value 3.14159 |
| Random | Provide random numbers from 0 through 1 |
| Read | Read variable |
| ReadLn | Read line of variables |
| Real | Define real variable |
| Reset | Open file for reading |
| Rewrite | Open file for writing |
| Sin | Sine |
| Sqr | Square |
| Sqrt | Square root |
| True | Value of Boolean variable |
| UpCase | Convert lowercase to uppercase |
| Write | Write variable |
| WriteLn | Write line of variables |

# Summary of
# Turbo Pascal

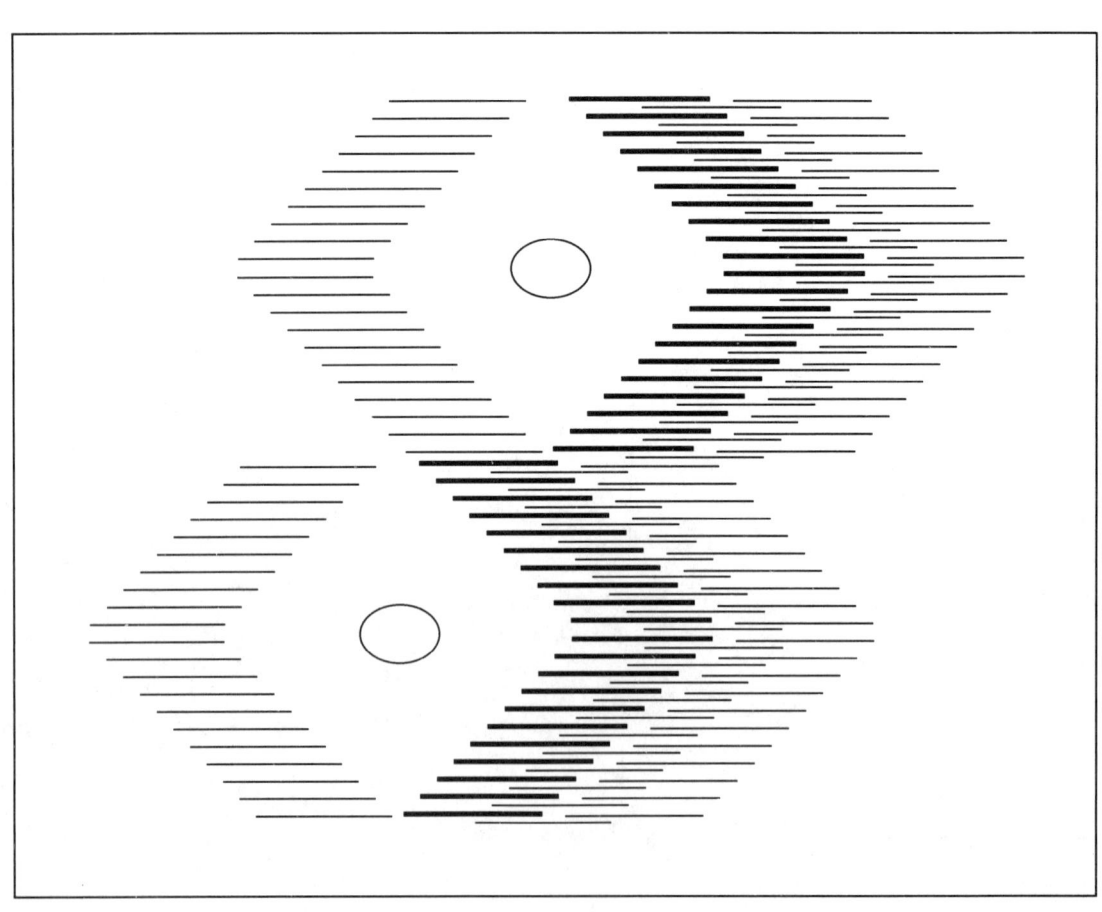

*Appendix* **B**

## Standard Character Set

The standard character set consists of the following:

| | |
|---|---|
| Alphabet | A-Z and a-z |
| Digits | 0-9 |
| Underline | _ |
| Blank character(space) | |
| Special characters | + - * / = < > ( ) |
| | [ ] { } . , : ; # ' " |

## Variable Names (Identifiers)

A Turbo Pascal variable name can be any sequence of alphabetic and numeric characters and the underline character, but the first character

cannot be a number. Uppercase and lowercase letters are the same. There may be as many as 127 characters in the name.
Examples:

    X   Mean   Sum   Sum_X   Y_Calc   Plot   Euler

# Numbers

Integers are signed or unsigned strings of digits that range from $-32768$ through $32767$.
Examples:

    15   $-19253$   0

Reals are written with a decimal point, a scale factor, or both.
Examples:

    0.5   0.57721566   3.7912E2   3.7912E-4

The E notation designates multiplication by a power of 10. Thus, E2 means 100. The E must be followed by an integer, signed or unsigned. The exponent ranges from $-38$ through 38 for Turbo Pascal and $-307$ through 308 for Turbo-87 Pascal.

# Comments

Comments are surrounded by the braces, { and }, and they are ignored by the compiler.
Example:

    { get values for N and arrays X and Y }

# Operations

Turbo Pascal performs integer, real, Boolean, and relational operations.

## Integer Operations

|     |     |
| --- | --- |
| $+$ | Addition |
| $-$ | Subtraction |
| $*$ | Multiplication |
| DIV | Division, truncated to an integer |
| MOD | Modulo supplies the remainder from the division of two integers |

## Real Operations

| | |
|---|---|
| + | Addition |
| − | Subtraction |
| * | Multiplication |
| / | Division |

## Boolean Operations

**AND   OR   NOT**

## Relational Operations

The following operations produce the Boolean values True or False:

| | |
|---|---|
| < | Less than |
| > | Greater than |
| − | Equal to |
| < = | Less than or equal to |
| > = | Greater than or equal to |
| <> | Not equal to |
| **IN** | Determine membership in a set |

# Syntax

The proper syntax must be used to define program headings, constants, variables, procedures, and functions and to make assignment and compound statements. The placement of semicolons is also important.

## Program Heading

The form for program headings is

**PROGRAM** Program_Name;

Example:

**PROGRAM** Least3;

## Constant Definition

A constant must be defined as a simple value, not as an expression.

The form for defining constants is

```
CONST
   Const_Name = value;
   Const_Name = value;
```

Example:

```
CONST
   Maxr = 20; { data points }
   Tol = 1.0E-5;
```

## Variable Definition

The form for defining variables is

```
VAR
   Var_Name, Var_Name: Type;
   Var_Name, Var_Name: Type;
```

Example:

```
VAR
   Error        : Boolean;
   Nrow, Ncol   : Integer;
   X, Y, Y_Calc: Real;
```

## Assignment Statements

The form for assigning values to variables is

```
Var_Name: = expression
```

Example:

```
New_Term := Sum;
Y := X + 2;
Gamma := Gam/(S * (X + 1));
```

## The Compound Statement

A compound statement is any sequence of statements enclosed by a **BEGIN END** pair.

Example:

```
BEGIN { swap }
  Hold := P;
  P := Q;
  Q := Hold
END; { swap }
```

## Procedure Definition

A procedure is a separate routine that begins with a header and contains a **BEGIN END** block.

```
PROCEDURE Proc_Name (Parameters; VAR variable parameters);
  body of procedure
```

Example:

```
PROCEDURE Deriv (X: Real; VAR Fx, Dfx: Real);

VAR
  E: Real;
BEGIN
  E := Exp(X);
  Fx := E − 4.0*X;
  Dfx := E − 4.0
END;
```

## Function Definition

A function is a separate routine that returns a value. It begins with a header and contains a **BEGIN END** block.

```
FUNCTION Func_Name (Parameters); Result_Type; body of function
```

Example:

```
FUNCTION Fx(X: Real): Real;
BEGIN
    Fx := 1.0/X
END;
```

## *Placement of Semicolons*

Semicolons are required between any two Turbo Pascal statements. They are not required either immediately after a **BEGIN** or immediately before an **END**, but their presence there is harmless. A semicolon must *never* immediately precede the **ELSE** in an **IF-THEN-ELSE** construction.

# *Conditional Statements*

Conditional statements are executed when the corresponding logical expression is true. Otherwise, the **ELSE** clause is executed if it is present.

## *The IF-THEN Statement*

The form of the **IF-THEN** statement is

**IF** logical expression **THEN** conditional statement

Examples:

```
IF B[N, N] = 0.0 THEN Error := True;
IF NOT Error THEN WriteLn('X = ', X);
```

## *The IF-THEN-ELSE Statement*

The form of the **IF-THEN-ELSE** statement is

```
IF logical expression THEN
    conditional statement
ELSE
    alternate statement;
```

Examples:

```
IF Det = 0.0 THEN
  BEGIN
    Error : = True;
    WriteLn('ERROR: Matrix singular')
  END
ELSE
  BEGIN
    Setup(B, Coef, 1);
    Setup(B, Coef, 2);
    Setup(B, Coef, 3)
  END;
```

## *Iterative Statements*

Turbo Pascal statements can be repeatedly executed with the following loop constructions.

### *The* **WHILE-DO** statement

The form of the **WHILE-DO** statement is

**WHILE** expression **DO** statement;

Example:

**WHILE** Pivot < A[J] **DO** J := J − 1;

### *The REPEAT-UNTIL statement*

The form of the **REPEAT-UNTIL** statement is

**REPEAT** statements **UNTIL** logical expression

Examples:

```
REPEAT
    Write('How many equations?');
    ReadLn(N)
UNTIL N < Maxr;
```

```
REPEAT
    UNTIL KeyPressed;
```

### *The FOR-TO-DO Statement*

The form of the **FOR-TO-DO** statement is

**FOR** variable := first-value **TO** last-value **DO** statement

Example:

**FOR** J := 1 **TO** N **DO** Write(Coef[J]);

### *The FOR-DOWNTO-DO Statement*

The form of the **FOR-DOWNTO-DO** statement is

**FOR** variable := first-value **DOWNTO** last-value **DO** statement

Example:

```
FOR I := Terms DOWNTO 1 DO
  BEGIN
    Sum := 1.0 + I * V/U;
    U := Sum
  END;
```

# Transfer-of-Control Statements

Sometimes it is necessary to prematurely terminate a loop or procedure. This is accomplished with the Turbo Pascal Exit and Halt commands. (The GOTO statement also performs this function, however, you should avoid using it for termination because it makes the program more difficult to understand.)

## The Exit Statement

The Turbo Pascal Exit statement causes the program to prematurely leave the current block or subroutine.

Example:

```
IF Index[K, 3] > 1 THEN
  BEGIN
    WriteLn(' ERROR: matrix singular');
    Error := True;
    Exit { procedure Gaussj }
  END;
```

## The Halt Statement

The Turbo Pascal Halt statement causes the program to prematurely terminate.

Example:

```
IF Error THEN Halt;
```

# Input and Output
## Input Procedures

Read(variables);                Gets values from the current line or record
                                and assigns them to the corresponding

variables. Unread data on a line are available to the next Read statement.

ReadLn(variables);

Gets values from the current line or record. Unread data on the line are ignored. The next Read or ReadLn begins at the next line or record.

Read(File_Var, variables);

Gets values from a disk file. File_Var is a symbol declared to be type **FILE** and is associated with the corresponding file name through an Assign statement. Unread data on the line in the file are available to the next Read statement.

ReadLn(File_Var, variables);

Gets values from a disk file. File_Var is as described above. Unread data on the line are ignored. The next Read or ReadLn statement begins at the next line or record in the file.

## Output Procedures

Write(variables);

Places the corresponding data into the current line for display on a video screen.

WriteLn(variables);

Appends its output to the current line or record, and then begins a new line.

Write(File_Var, variables);

Places the corresponding data into the current line to be written to a disk file. File_Var is a symbol declared to be type **FILE** and associated with the corresponding file name through an Assign command.

WriteLn(File_Var, variables);

Appends its output to the current line or record in a disk file, and then begins a new line. File_Var is as described above.

## Formatting Numeric Output

Turbo Pascal uses a standard form to display numeric expressions through a Write or WriteLn statement. For example, floating-point numbers have 11 digits and a 2-digit exponent. Turbo-87 Pascal displays 15 digits and a 3-digit exponent. The appearance of a number can be changed by appending one or two integers to the expression, as follows:

| | |
|---|---|
| integer_expression: integer; | Specifies the field width. If the field is wider than the number, spaces are placed on the left side. If the specified field width is too small, it is automatically enlarged to the necessary width. |
| real_expression:integer; | Includes the exponent. |
| real_expression:integer: integer; | The second integer specifies the number of digits in the mantissa. No exponent is given. |

Example:

```
WriteLn(I:3, X[I]:8:1, Y[I]:9:2, Y_Calc[I]:11)
```

# Data Types

**TYPE** declarations follow **CONST** declarations and precede **VAR** declarations.

## Scalar Types

The form for scalar types is

```
TYPE
   Type_Name (identifier, identifier);
```

Example:

```
TYPE
   Day_of_week = (Monday, Tuesday, Wednesday,
                  Thursday, Friday,
                  Saturday, Sunday);
```

## Subrange Types

The forms for subrange types are

```
TYPE
   Type_Name = constant, constant;
VAR
   Var_Names: constant, constant;
```

Example:

**TYPE**
Index = 1..Max;
Week_Day = (Monday);

## *Array Types*

The forms for array types are

**TYPE**
Type_Name = **ARRAY**[index_type, index_type] **OF** type;
**VAR**
Var_Names: **ARRAY**[index_type, index_type] **OF** type;

An array with more than one dimension requires that an index_type be specified for each dimension. An index_type is usually a subrange type, but it can be any scalar type.

Examples:

**TYPE**
Arys  = **ARRAY**[1..Cmax] **OF** Real;
Ary2s = **ARRAY**[1..Rmax, 1..Cmax] **OF** Real;
**VAR**
W  : **ARRAY**[1..Maxc, 1..Maxc] **OF** Real;
Index: **ARRAY**[1..Maxc, 1..3] **OF** Integer;

## *Referencing Array Elements*

Array indexes appear between brackets. For multidimensional arrays, the index expressions are separated by commas. The subscript index for each dimension can be any expression that evaluates to the declared index type.

Array_Name[index, index];

Example:

C[I, J] := A[I, K] * B[K, J];

## Set Types

The forms for set types are

```
TYPE
   Type_Name = SET OF base_type;
VAR
   Var_Type: SET OF base_type;
```

Example:

```
TYPE
   Day_of_week = (Monday, Tuesday, Wednesday,
                        Thursday, Friday,
                        Saturday, Sunday);
   Week = SET OF Day_of_Week;
VAR
   Workday, Holiday, Weekday: Week;
```

## Set Operations

The following are set operations:

| | |
|---|---|
| + | Union |
| * | Intersection |
| − | Difference |
| = | Set equality |
| <> | Set inequality |
| **IN** | Set membership. True if all the scalar operand on the left is a member of the set operand on the right |
| < = | Set inclusion. True if all elements of the left operand are also part of the right operand |
| > = | Set containment. True if all elements of the right operand are also part of the left operand |

## File Types

Before a disk file can be referenced, an association must be made between the file name and a Turbo Pascal symbol. This association is accomplished with the Assign statement.

Example:

**VAR**
  Filename: **STRING**[14];
  Filvar: Text;

The following Turbo Pascal input and output routines are used with files:

| | |
|---|---|
| Assign | Associate file name with identifier. |
| Reset | Open a file for reading. |
| Rewrite | Open a file for writing. |
| Eof | Indicate when the end of a file has been reached (a Boolean function). |
| Close | Release the file handle for a file that is no longer needed (needed to write the directory entry for a new file). |

# Answers to
# Exercises

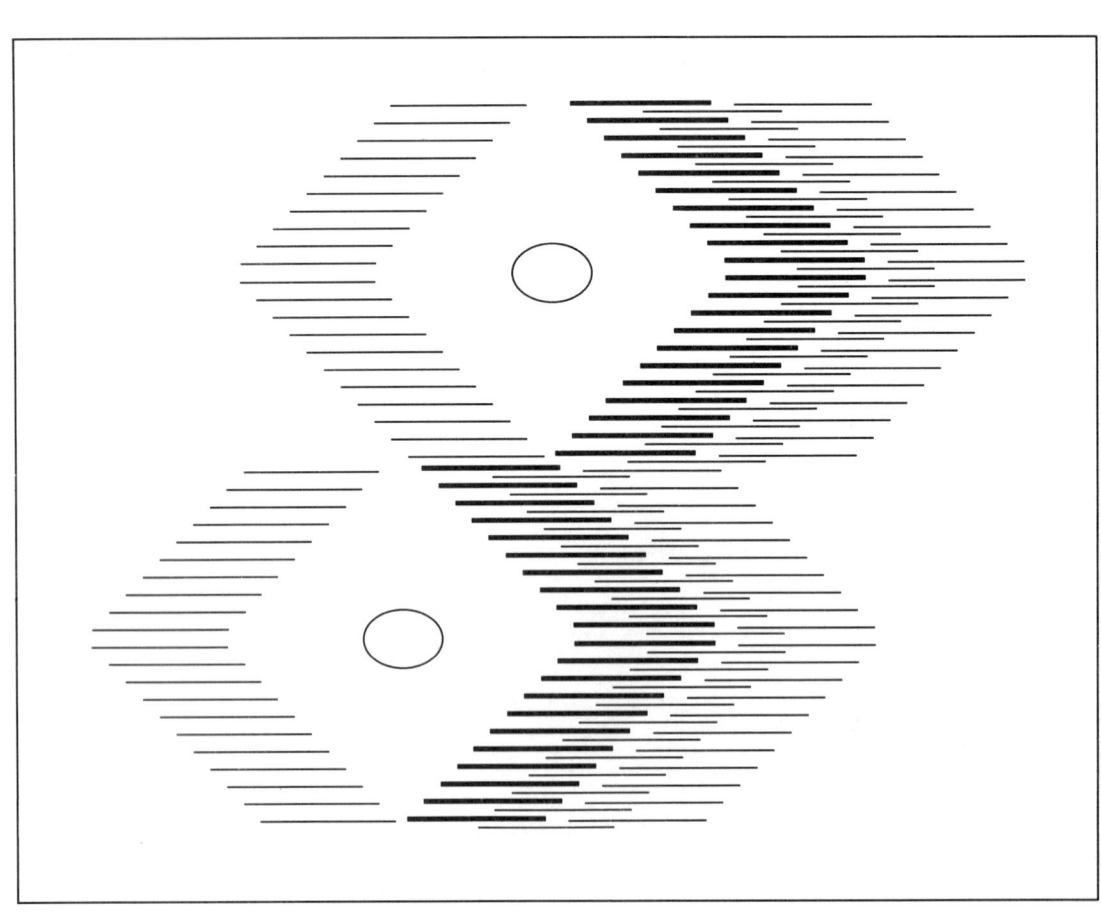

*Appendix* **C**

**4.3.** $I_1 = 2.39 + j1.71 = 2.9/35.5$
$I_2 = 1.87 + j0.117 = 1.87/3.58$

**5.1.** $A = 20.46, B = -5153$

**7.1.** $A = -2.128, B = 2.319, C = -0.548$
**7.2.** $A = 20.57, B = -7469, C = -0.323$
**7.3.** $A = 3.37, B = 0.00198, C = -177100$

**8.1.** $X = 0.758$
**8.2.** 30.8 liters

**9.1.** 5.938; the Romberg method
**9.2.** 0.8220; Simpson's method
**9.3.** 0.785; the trapezoid method
**9.4.** 0.785; the trapezoid method

**10.1.** 0.6 ev.
**10.2.** $Q = 15.2$ Kcal/mole, $A = 4470$/min.

# Program Index

Adding a call to procedure Plot, 124
The arc cosine function, 9
An arc sine function, 8
An arc tangent function with two arguments, 7
The beginning of a curve-fitting program, 116–117
Bessel functions of the first kind, 287–289
Bessel functions of the second kind, 291–292
The best fit to a set of linear equations, 94–95
A bubble sort routine, 148
Calculating the Gaussian error function with Simpson's method, 274–275
Calculating the mean and standard deviation, 21
Calculation of the Gamma function, 284–285
The complete curve-fitting program, 135–137
The current main program, 127
The determinant of a 3-by-3 matrix, 45–46
An equation of state for steam, 189–191
The error function and its complement, 279–280
Fitting the Clausing factor to the ratio of two polynomials, 249–251
A function to generate Gaussian distributed random numbers, 26
The Gauss-Jordan elimination procedure, 72–74
An improved trapezoidal method, 227–228
An infinite series expansion for the Gaussian error function, 276
Inputting the polynomial order from the keyboard, 176–178
A least-squares fit to the linearized exponential function, 253–255
A least-squares fit to the nonlinearized exponential function, 259–263
The main program for the second version of Newton's method, 208
Matrix multiplication program, 41–42
Newton's method with a loop counter, 213
Newton's method with a test for zero slope, 211
Newton's method, version one, 206
Numerical integration by the Romberg method, 235–237

Numerical integration by Simpson's method, 230–231
Numerical integration by the trapezoidal method, 226
A parabolic least-squares fit, 163–166
A parabolic least-squares fit using Gauss-Jordan elimination, 171–173
A plotting procedure, 121–123
Procedure Func for $e^x = 4x$, 214
Procedure Linfit to generate a least-squares fit, 131
Procedure Linfit to simulate a linear fit, 126
Procedure MeanStd to calculate the mean and standard deviation, 20
Procedure Square, 43
Procedure Swap to interchange values, 147
Procedures Get_Data and Linfit for the heat-capacity equation, 181–182
Procedures Get_Data and Linfit for the vapor pressure equation, 184–185
Program to sort alphabetic data on disk, 155–156
Program to test the sorting routines, 145–146
Programming the Romberg method with adjustable panels, 241–242
Reading the data from a disk file, 89–91
A recursive quick sort, 151–152
The revised procedure Write_Data, 126
A Shell sort routine, 149–151
Simultaneous solution of equations with complex coefficients, 101–103
Simultaneous solution by Gauss elimination, 64–66
Solution of linear equations by the Gauss-Seidel method, 108–111
Solution of a set of ill-conditioned equations, 86–87
Solution of $Sin(X) = X/10$, 215
Solution of simultaneous equations by Gauss-Jordan elimination, 70–72
Solution of simultaneous equations and matrix inversion by the Gauss-Jordan elimination method, 78–82
Solution of three linear equations by Cramer's rule, 57–59
Solution of the vapor pressure equation, 217
Testing the arc sine and arc tangent functions, 9
Testing the built-in random-number generator, 24
Testing floating-point operations, 3
Testing the Gaussian random-number generator, 27
Testing the Ln and Exp functions, 6

# Subject Index

**A**

Activation energy, 253, 265
Activity coefficient, 195
Arc cosine function, 8–9
Arc sine function, 8–9
Arc tangent function, 7–9
Arrays, 31
  as procedure parameter, 21–22
Asymptotic expansion, 278, 290
Average, 15

**B**

Back substitution, 62
Band gap, 264
Bell-shaped curve, 17
Bessel
  equation, 282, 284, 286
  function,
    first kind, 286–290
      plot of, 287
      selected values of, 289
    second kind, 290–293
      plot of, 293
      selected values of, 293
Best-fit solution, 92–97
Boltzmann constant, 264
Bubble sort, 147–148

**C**

Clausing factors, 248–252
Coefficient matrix, 53
Color, selecting, 2
Column
  index, 36
  vector, 33
Complementary error function,
    278–281
Complex roots, 97–105
Conformable matrix, 39
Convergence
  absolute criterion for, 107
  lack of, 212–214
  to a root, 200–205
Coprocessor, 4
Correlation coefficient, 132–138
Cramer's rule, 53, 57–60
Cross product, 34
Crystallization, 265
Cumulative distribution function,
    268–269, 277
Curve fitting,
    (see least-squares curve fitting)

**D**

Data plotting, 120–128
Determinant, 44–46, 85

Diffusion, 253, 271–273, 277
Diffusion coefficient, 253, 271–272
Diffusivity, 271
Disk file,
    reading coefficients from, 89–92
    sorting from, 155–157
Dispersion, 16–18
Dissociation, 217
Distribution function, 268–269,
        277
Dot product, 34
Dynamic range, floating–point, 2

**E**

EGA, 3
Electrical circuit equation,
    dc, 55–57
    ac, 98–105
Elimination,
    Gauss, 60–68
    Gauss–Jordan, 68–83
Equation of state, 186–194, 217
Error function, 271–278
    Selected values, 278
Error function complement,
        278–281
    Selected values, 281
Exponential curve fit, 253–264
Exponential equation,
    direct solution of, 256–263
    linearization of, 252–256
External files, 10–11

**F**

Factorials, 282, 284
Fick's
    first law, 271
    second law, 272
Floating-point operations
    dynamic range, 2

Floating-point operations (cont.)
    precision, 2
    roundoff error, 2–4
    test of, 3–5
Function plotting, 120–128

**G**

Gamma function, 281–285, 286
    plot of, 283
    selected values, 285
Gauss elimination, 60–68
Gauss-Jordan elimination, 68–83
Gauss-Seidel
    iteration, 105–112
    point relation, 107
Gaussian
    distribution, 17
    error function, 271–278
    error function complement,
        278–281
    random numbers, 25–28
Graphite heat capacity, 196

**H**

Heat capacity,
    equation, 180–183
    of graphite, 196–197
Hilbert matrix, 83–89
Hydrogen sulfide dissociation, 217

**I**

Ill-conditioned matrices, 83–92,
        174
INCLUDE directive, 10–11
INCLUDE files,
    ASIN, 8
    ATAN, 7
    GAUSSJ, 72–74
    MEANSTD, 20

INCLUDE files (cont.)
   PLOT, 121–123
   RANDG, 26
   SORT,
      bubble, 147–148
      Shell, 148–150
      quick, 150–152
   SQUARE, 43
   SWAP, 147
Infinite series, 275
Integration, numerical, 221–244
   adjustable panels, 240
   of periodic function, 232–234, 239
   infinite at one limit, 240
   the Romberg method, 234–244
   Simpson's method, 229–234
   the trapezoidal method, 223–229
Inverse
   cosine, 8–9
   sine, 8–9
   tangent, 7–9
Inversion, matrix, 46–47, 68
Iteration,
   Gauss-Seidel, 105–112
   Newton's method, 188–218

**K**

Kirchhoff voltage law, 56, 98

**L**

Least-squares curve fitting, 115–138, 161–194
   adjusting order of polynomial, 175–180
   data matrix, 169
   data vector, 169
   derivation, 128–131, 162

Least-squares curve fitting (cont.)
   heat capacity, 180–183
   nonlinear, 186–194, 247–265
   parabolic, 162–168
   polynomial, 162
   rational function, 248–252
   superheated steam, 186–194
   three variables, 186
   vapor pressure, 139, 183–186, 196
Linear
   dependence, 54–55
   equation, 52
Linearization, 247, 252
Loop currents, 56

**M**

Magnitude of vector, 33
Math coprocessor, 4
Matrix, 36–47
   addition, 38
   coefficient, 53
   column index, 36
   conformable, 39
   determinant, 44–46, 85
   diagonal, 37
   division, 46–47
   identity, 45
   ill-conditioned, 83–89, 174
   inversion of, 46–47, 83, 174
   main diagonal, 37
   multiplication, 39–42, 174
   principal diagonal, 37
   row index, 36
   scalar multiplication, 38
   singular, 46, 54
   square, 37
   subtraction, 38
   symmetric, 38
   transpose, 37, 39, 170
   unit, 37

Mean value, 15
Mercury vapor pressure, 196
Multiple constant vectors, 75–83

**N**

Newton's method, 199–218, 258
    multiple roots, 215
    no convergence, 212–214
    nonlinear curve fitting,
        247–255
    test for zero slope, 209–212
Newton-Raphson method,
    199–218
Nonlinear equation, 200, 214,
    247–265
Nonpolynomial equation, 180–194
Normal distribution, 17, 268
Normally distributed random
    numbers, 25–28
Numerical quadrature
    (see integration)

**P**

Parameter, procedure as, 207
Parabolic
    curve fitting, 162–168
    equations, 162
Pascal, summary of, 301–313
Phasor angle, 100
Pivot element, 61–63, 106–107
Plot routine, 120–128
Point relaxation, 107, 112
Polynomial equations, 162
Precision, floating-point, 2
Principal diagonal, 37
Probability density function, 268

**Q**

Quick sort, 150–152

**R**

Random numbers, 23–28
    evaluation of, 23
    from $\pi$, 25
    Gaussian, 25–28
    uniform, 23–24
Rational function, 248–252
Recursion,
    gamma function, 282
    quick sort, 150–152
Regression (see least-squares)
Relaxation, Gauss-Seidel, 105–112
Residual, 128–129
Reserved words, 297
Romberg method, 234–244
Root finding, 199–218
Row index, 36
Row vector, 32

**S**

Scalar variable, 32
SEE (Standard Error of the
        Estimate), 134, 173–174
Shell-Metzner sort, 148–150
Simpson's rule, 229–234, 273
Simultaneous best fit, 92–97
Simultaneous solution, 51–117
    back substitution, 62
    best fit, 92–97
    complex coefficients, 97–105
    data on disk, 89–92
    Gauss elimination, 60–68
    Gauss-Jordan elimination,
        68–83
    Gauss-Seidel, 105–112
    Hilbert matrices, 83–89
    multiple vectors, 75–83
    nonlinear equations, 105–112
Singular matrix, 46, 54

Sorting routines, 143–157
  bubble, 147–148
  quick sort, 150–152
  Shell-Metzner, 148–150
SRS (sum of residuals squared),
  128–129, 256–257
Standard deviation, 18–22,
  269–271
Standard error, 134, 174
Standard identifiers, 298
Stirling's approximation, 282
String variable, 35
Superheated steam, 186–194
Swap routine, 147
Symmetric matrix, 38

**T**

Tolerance, 2, 206–207, 233
Transpose, matrix, 37, 39
Trapezoidal method, 223–229

**U**

Unit matrix, 37

**V**

Van der Waals equation, 217
Vapor pressure equation, 183–186,
  200, 216
  of mercury, 196
  of water, 139
Vector, 32–36
  addition, 34
  column, 33
  cross product, 34
  dot product, 34
  magnitude, 33
  multiplication, 33
  row, 32
  scalar product, 34
  strings, 35
  vector product, 34

**Z**

Zero slope, 209
Zinc, diffusion in copper, 253, 277

# Selections from The SYBEX Library

## Languages

### BASIC

**BASIC PROGRAMS FOR SCIENTISTS AND ENGINEERS**
**by Alan R. Miller**
318 pp., illustr., Ref. 073-8
A course in mathematical problem-solving and BASIC programming techniques, as well as a sourcebook of practical, ready-to-use programs for statistical analysis, linear and non-linear curve-fitting, numerical integration and more.

### PASCAL

**INTRODUCTION TO TURBO PASCAL**
**by Douglas S. Stivison**
268 pp., illustr., Ref. 269-8
This bestseller introduces Pascal programming in the environment of Turbo Pascal, giving realistic examples from the author's programming experience. The focus is on how to get all the benefits offered by this Pascal implementation.

**INTRODUCTION TO PASCAL, INCLUDING TURBO PASCAL**
**by Rodnay Zaks**
464 pp., illustr., Ref. 319-8
The latest edition of the SYBEX classic that has trained thousands and earned wide-spread praise. The book's clear, systematic approach covers every featureof Pascal, with exercises and answers in both ISO Standard and Turbo Pascal.

**TURBO PASCAL LIBRARY**
**by Douglas S. Stivison**
221 pp., illustr., Ref. 330-9
More than a collection of time-saving, problem-solving Turbo routines, this is a sourcebook on good programming style. With programs for games, system utilities, business, and engineering.

**INTRODUCTION TO THE UCSD P-SYSTEM**
**by Charles W. Grant and Jon Butah**
300 pp., illustr., Ref. 061-X
"Outstanding, easy to use . . . thorough treatment . . . Excellent as an introduction for newcomers, or as a reference for experienced programmers."
—*Computer Book Review*

**INTRODUCTION TO PASCAL (Including UCSD Pascal)**
**by Rodnay Zaks**
420 pp., 130 illustr., Ref. 066-0
A step-by-step introduction for anyone who wants to learn the Pascal language, describing UCSD and Standard Pascals. No technical background is assumed.

**THE PASCAL HANDBOOK**
**by Jacques Tiberghien**
486 pp., 270 illustr., Ref. 053-9
A dictionary of the Pascal language, defining every reserved word, operator, procedure, and function found in all major versions of Pascal.

**FIFTY PASCAL PROGRAMS**
**by Bruce H. Hunter**
338 pp., illustr., Ref. 110-1
More than just a collection of useful programs! Structured programming techniques are emphasized and concepts such as data type creation and array manipulation are clearly illustrated.

## THE C LANGUAGE

### PROGRAMMING THE MACINTOSH IN C
**by Bryan J. Cummings and Lawrence J. Pollack**
294 pp., illustr., Ref. 328-7
A complete introduction to the C language and structured programming using Consulair C claimed to be the number one Mac-resident C compiler. Includes sample programs, a glossary, and a reference guide.

### UNDERSTANDING C
**by Bruce H. Hunter**
320 pp., Ref. 123-3
Explains how to program in powerful C language for a variety of applications. Some programming experience assumed.

### MASTERING C
**by Craig Bolon**
400 pp., illustr., Ref. 326-0
Designed for the programming professional, this gives a complete description of C language programming, focusing on how to get the most power, efficiency, and portability out of C.

### DATA HANDLING UTILITIES IN C
**by Robert Radcliffe and Thomas Raab**
500 pp., illustr., Ref. 304-X
This is a "Software Toolkit" of useful C functions, techniques and usable code for commercial application programmers and software developers. Because commercial programs require high user-interaction and permanent files, the book concentrates on data entry, validation, display, and efficient data storage. There is a comprehensive section all about logical data types and another giving sample applications.

# Technical

## ASSEMBLY LANGUAGE

### ASSEMBLY LANGUAGE TECHNIQUES FOR THE IBM PC
**by Alan Miller**
350 pp., illustr., Ref. 309-0
Any IBM PC user and programmer that wants to learn techniques to get more power from the PC will find the tutorial and program library elements in this title extremely valuable. Programs included in the book allow the reader to do such tasks as transferring WordStar to ASCII and back, switch from color screens to monochrome screens and back, set the printer to any typeface, and more. Techniques are given for the programmer to generate more programs.

### PROGRAMMING THE 6502
**by Rodnay Zaks**
408 pp., illustr., Ref. 135-7
A comprehensive introduction to 6502 assembly language, from elementary concepts to advanced data structures and program development. This bestseller is now in its fourth edition.

### PROGRAMMING THE 65816
**by William Labiak**
370 pp., illustr., Ref. 324-4
A complete introduction and indispensable reference for programmers of any member of this family of microprocessors. Topics include 6502 emulation, 8- and 16-bit modes, and much more.

### PROGRAMMING THE 8086/8088
**by James W. Coffron**
311 pp., illustr., Ref. 120-9
Assembly language programming for power and performance on the IBM PC and other 8086/8088-based systems—from basic concepts to advanced techniques. Sample code and complete reference material.

### PROGRAMMING THE 68000
**by Steve Williams**

539 pp., illustr., Ref. 133-0

A step-by-step introduction to assembly language programming for 68000-based systems—from the basics of microprocessor operation to I/O programming. An excellent programmer's handbook.

### PROGRAMMING THE Z80
### (3rd Edition)
**by Rodnay Zaks**

624 pp., illustr., Ref. 069-5

Let the reviewers speak for this educational classic: "Exceptionally well organized and well written . . . recommended for individuals with a specific interest in the Z80 as well as persons with a general interest in computer programming."

*—JEPT*

### PROGRAMMING THE APPLE II
### IN ASSEMBLY LANGUAGE
**by Rodnay Zaks**

519 pp., illustr., Ref. 290-6

All elements of the art of assembly language programming for the current Apple IIc and Apple IIe are covered in Zaks' classic style.

### PROGRAMMING THE
### MACINTOSH IN ASSEMBLY
### LANGUAGE
**by Steve Williams**

400 pp., illustr., Ref. 263-9

This is an up-to-date tutorial and reference guide to programming the 68000 in the Macintosh environment. Covering architecture, instruction set, Toolbox, and advanced programming concepts, this is ideal for intermediate to professional applications programmers.

## HARDWARE

### MASTERING DIGITAL DEVICE
### CONTROL
**by William G. Houghton**

350 pp., illustr., Ref. 346-5

A clear, general methodology for system design using single-chip microcontrollers. Topics include memory and I/O, display devices, and interfacing with multi-chip CPUs. With practical illustrations.

### MICROPROCESSOR
### INTERFACING TECHNIQUES
**by Rodnay Zaks and Austin Lesea**

456 pp., 400 illustr., Ref. 029-6

Complete hardware and software interfacing techniques, including D to A conversion, peripherals, bus standards and troubleshooting.

### THE RS-232 SOLUTION
**by Joe Campbell**

194 pp., illustr., Ref. 140-3

Finally, a book that will show you how to correctly interface your computer to any RS-232-C peripheral.

### MASTERING SERIAL
### COMMUNICATIONS
**by Joe Campbell**

250 pp., illustr., Ref. 180-2

A complete guide to the software aspects of serial communications, including hard-to-find details on the IBM PC's serial programming and the Kermit and XMODEM protocols. Sample programs in C, BASIC, and assembly language.

## OPERATING SYSTEMS

### PROGRAMMERS'S GUIDE
### TO GEM
**by Philip Balma and William Fitler**

504 pp., illustr., Ref. 297-3

A detailed handbook on programming in the GEM environment, stressing quality applications-building using GEM tools. With many programming examples, and specific information on the Atari St, IBM PC, and other GEM-compatible systems.

### PROGRAMMER'S GUIDE TO
### WINDOWS
**by David Durant**

450 pp., illustr., Ref. 362-7

An indispensable guide to programming under Microsoft Windows—including techniques for portable programming and using the Software Development Kit. Examples in C, Pascal, and assembly language.

## REAL WORLD UNIX

**by John D. Halamka**

209 pp., Ref. 093-8

This book is written for the beginning and intermediate UNIX user in a practical, straightforward manner, with specific instructions given for many business applications.

## THE PROGRAMMER'S GUIDE TO TOPVIEW

**by Alan R. Miller**

313 pp., illustr., Ref. 273-6

This guides the programmer through all the major features of TopView for the entire IBM PC line. This includes examples of programs on TopView, descriptions of subroutine calls and macros, and instructions for writing including assembly language programming. Special emphasis is given to writing programs that run both with or without TopView.

# Introduction to Computers

## THE SYBEX PERSONAL COMPUTER DICTIONARY

120 pp. Ref. 199-3

All the definitions and acronyms of micro computer jargon defined in a handy pocket-sized edition. Includes translations of the most popular terms into ten languages.

## FROM CHIPS TO SYSTEMS: AN INTRODUCTION TO MICROPROCESSORS

**by Rodnay Zaks**

552 pp., 400 illustr., Ref. 063-6

A comprehensive introduction to microprocessors from both a hardware and software standpoint: what they are, how they operate, how to assemble them into a complete system.

# Special Interest

## CELESTIAL BASIC

**by Eric Burgess**

300 pp. 65 illustr. Ref. 087-3

A collection of BASIC programs that rapidly complete the chores of typical astronomical computations. It's like having a planetarium in your own home! Displays apparent movement of stars, planets and meteor showers.

## PERSONAL COMPUTERS AND SPECIAL NEEDS

**by Frank G. Bowe**

175 pp., illustr. Ref. 193-4

Learn how people are overcoming problems with hearing, vision, mobility, and learning, through the use of computer technology.

# Computer Specific

## *AMIGA*

### AMIGA PROGRAMMER'S HANDBOOK Volume 1

**by Eugene Mortimore**

575 pp., illustr., Ref. 367-8

All the Amiga's power at your fingertips! Organized for working programmers, this is an A to Z compendium of Amiga system facilities, including ROM-BIOS exec calls, the Graphics Library, Animation Library, Layers Library, Intuition calls, and the Workbench.

## *APPLE II - MACINTOSH*

### PROGRAMMING THE MACINTOSH IN ASSEMBLY LANGUAGE

**by Steve Williams**

400 pp., illustr. Ref. 263-9

Information, examples, and guidelines for programming the 68000 microprocessor are given, including details of its entire instruction set.

## THE PRO-DOS HANDBOOK
### by Timothy Rice and Karen Rice
225 pp., illustr. Ref. 230-2

All Pro-DOS users, from beginning to advanced, will find this book packed with vital information. The book covers the basics, and then addresses itself to the Apple II user who needs to interface with Pro-DOS when programming in BASIC. Learn how Pro-DOS uses memory, and how it handles text files, binary files, graphics and sound. Includes a chapter on machine language programming.

## USING THE MACINTOSH TOOLBOX WITH C
### by Fred A. Huxham, David Burnard and Jim Takatsuka
559 pp., illustr., Ref. 249-3

In one place, all you need to get applications runnning on the Macintosh, given clearly, completely, and understandably. Featuring the C language.

## MASTERING Pro-DOS
### by Timothy Rice and Karen Rice
250 pp., illustr., Ref. 315-5

This companion volume to The ProDOS Handbook contains numerous examples of programming techniques and utilities that will be valuable to intermediate and advanced users.

## THE EASY GUIDE TO YOUR MACINTOSH
### By Joseph Caggiano
214 pp., illustr., Ref. 216-7

Simple and quick to use, this tells first time users how to set up their Macintosh computers and how to use the major features and software.

## MACINTOSH FOR COLLEGE STUDENTS
### by Bryan Pfaffenberger
250 pp., illustr., Ref. 227-2

Find out how to give yourself an edge in the race to get papers in on time and prepare for exams. This book covers everything you need to know about how to use the Macintosh for college study.

# ATARI

## UNDERSTANDING ATARI ST BASIC PROGRAMMING
### by Tim Knight
300 pp., illustr., Ref. 344-9

Here is a comprehensive tutorial and reference guide for ATARI ST BASIC programming, including graphics, sound and GEM windows. With a complete ST BASIC command summary.

# CP/M SYSTEMS

## THE CP/M HANDBOOK
### by Rodnay Zaks
320 pp., illustr., Ref 048-2

An indispensable reference and guide to CP/M – complete in reference form. "An excellent reference guide . . ." Dr. Dobbs Journal

## MASTERING CP/M
### by Alan Miller
398 pp., illustr., Ref. 068-7

For advanced CP/M users or systems programmers who want maximum use of the CP/M operating system: this book takes up where the CP/M Handbook leaves off.

## THE CP/M PLUS HANDBOOK
### by Alan Miller
250 pp., illustr., Ref. 158-6

This guide is easy for beginners to understand, yet contains valuable information for advanced users of CP/M Plus.

## MASTERING DISK OPERATIONS ON THE COMMODORE 128
### by Aian R. Miller
f238 pp., illustr., Ref. 357-0

This guide to using CP/M Plus on the Commodore 128 is essential for users at all levels, offering introductory tutorials, in depth treatment of major topics, a loo inside the operating system, and a CP/M Plus command summary.

## SYBEX Computer Books
## are different.

# Here is why . . .

At SYBEX, each book is designed with you in mind. Every manuscript is carefully selected and supervised by our editors, who are themselves computer experts. We publish the best authors, whose technical expertise is matched by an ability to write clearly and to communicate effectively. Programs are thoroughly tested for accuracy by our technical staff. Our computerized production department goes to great lengths to make sure that each book is well-designed.

In the pursuit of timeliness, SYBEX has achieved many publishing firsts. SYBEX was among the first to integrate personal computers used by authors and staff into the publishing process. SYBEX was the first to publish books on the CP/M operating system, microprocessor interfacing techniques, word processing, and many more topics.

Expertise in computers and dedication to the highest quality product have made SYBEX a world leader in computer book publishing. Translated into fourteen languages, SYBEX books have helped millions of people around the world to get the most from their computers. We hope have helped you, too.

## mplete catalog of our publications:

2021 Challenger Drive, #100, Alameda, CA 94501
8233/(800) 227-2346   Telex: 336311

SYBEX, Inc. 2021 Challenger Drive, #100, Alameda, CA 94501
Tel: (415) 523-8233/(800) 227-2346  Telex: 336311